Rolling stoc… DEPAR… STOCK

Colin J. Marsden

LONDON

IAN ALLAN LTD

Front cover top: Plasser & Theurer 12 ton telescopic jib permanent way crane, classified under the TOPS computer classification system as YOB. *Plasser & Theurer*

Front cover bottom: 20.5 tonner centre chute fitted ballast hopper of the *Herring* or ZLV type displaying 'Isle of Wight' insignia at Sandown during 1982. *Colin J. Marsden*

First published 1984

ISBN 0 7110 1446 9

Published by Ian Allan Ltd, Shepperton, Surrey; and printed by Ian Allan Printing Ltd at their works at Coombelands in Runnymede, England.

Above: Class 08 No 08.501 hauls a short departmental train out of York CCE yard on 6 July 1983. *Colin J. Marsden*

Introduction

This, the third in the *Rolling Stock Recognition* series, deals with the departmental fleet. This ominous title covers a wide variety of railborne items, not only wagons but locomotives. EMUs, DMUs, engineers plant and 'Internal User' vehicles. Casual observers of the railway scene may classify all wagon stock under the same heading, however just over 25% of the freight stock in service is operated under the departmental flag. Usually stock taken over for departmental or internal operation is from the revenue earning fleets, but a small number of wagons and all track machines are purpose built.

To ease the recognition of individual vehicles the BR TOPS wagon code system has been followed for the freight stock, this conforming to codes Q, Y and Z. Locomotives are classified 97, and EMUs 92x or 93x. Track maintenance machines and Internal User vehicles are given separately. Each type or group has its own section headed where practicable by a list of types in service (these lists can only be taken as a guide as alterations are made on a near daily basis). It should be mentioned that a number of older wagons, particularly those which are unfitted, or fitted with vacuum brakes, will be withdrawn in the foreseeable future, as modern air braked stock becomes available.

The numbering of departmental stock is not as complex as it may at first seem. Vehicles of Group Q (former passenger/van stock) are usually in the following ranges DB975xxx, DB977xxx — vehicles taken into departmental stock from late 1966. Prior to this date departmental coaching stock was numbered in regional series: SR vehicles — DS70xxx, WR vehicles — DW150xxx, LMR vehicles — DM395xxx, ER vehicles — DE320xxx. However there are many anomalies to these series as will be found throughout the book. Freight stock of Groups Y and Z usually retain their former identities if converted from ex-revenue stock, with just the addition of the departmental number prefix. Purpose built departmental stock usually numbered in the 9xxxxx ranges carry their operator's prefix to the DB code. The prefix letters to departmental numbers indicate the following: all stock will have either a two or three letter prefix, the penultimate letter always a 'D' stands for departmental. This is either followed by a regional letter: S, W, E, M or for BR rebuilds post 1966, the letter B. Some modern air-braked civil engineers stock use a 'C' as the last letter indicating the Civil Engineering Department. The suffix letter indicates the owning body of the vehicle. CCE stock carry no suffix letter but other departmental codes are: A — CM&EE, C — BREL, K — S&T, L — CM&EE Electrification Unit, P — Shipping & International Services, R — Research Department, Derby, T — Traffic, X — Stores, Z — Public Affairs or Publicity, therefore vehicle ADB975032 is a post-1966 conversion operated by the CM&EE, while DW150352 is WR departmental used by the CCE.

It is hoped this product will assist readers in identifying some of the many departmental and service vehicles seen throughout BR. In no way can it set out to record every vehicle, but as many types as space has permitted have been included. Locations of vehicles will vary from time to time depending on their use, as will the careers of some stock, particularly those operated by the Research & Development section at Derby.

I would like to take this opportunity to record my thanks to many people who have given assistance in the production of this book and in particular the Public Affairs office at Waterloo, the BRB MEE Department, Mr J. Cormack of the Research & Development Division, Derby, and Mr Hans Mueller of Plasser & Theurer Ltd.

Colin J. Marsden
Surbition
August 1984

Below: This interesting cast plate was found attached to a bolster wagon during 1983, showing that the vehicle was once owned by the War Department as No 74, and registered with the LMS in 1941 for main line operation. *Colin J. Marsden*

Total Operations Processing System (TOPS)

This book is broadly based on the BR TOPS three letter vehicle code system, therefore a brief description is given. TOPS stands for *Total Operations Processing System*, a computer based communications and information storage system. The TOPS system for traffic control is a development of an American method introduced in the early 1960s for the Southern Pacific Transportation Company. In the closing years of the 1960s BR were actively involved in developing a computer based system for wagon and traffic control to ensure that (a) a constant supply of vehicles were available to BR and its customers, (b) a trace could be made at any time to ascertain the whereabouts of a vehicle and the nature of its loading, and (c) maintenance cycles for stock and locomotives could be closely monitored and adhered to. BR were faced with a decision of what type of system to adopt, whether to develop a completely new method or to utilise experience gained by the US network. Eventually a modified development of the Southern Pacific computer was decided upon and given the same name as its American counterpart — TOPS.

The BR TOPS master computer is situated in London and holds records of all BR locomotives, goods vehicles owned by BR or private companies, departmental stock and all passenger and NPCCS vehicles; in other words all BR/Private railborne vehicles can be traced by any TOPS terminal. Each locomotive or vehicle has a permanent entry giving basic information such as size, weight, capacity and maintenance. This is joined by 'active' information about the present status of each vehicle, this being fed into the master computer from any one of a number of local offices with TOPS reporting facilities. To differentiate the vast number of wagon types in operation a three letter code system was developed: the first letter indicating the wagon or type group, this is often derived by the type of vehicle ie: Group H is for hopper wagons, Group F for flat wagons. However there are a number of exceptions to this rule where more than one group is used for the same traffic purpose. Departmental vehicles only use the codes Q, Y and Z, together with Class 97 in the locomotive numbering system and 9xx in the EMU codes. The second letter of the three letter code identifies the actual type or section within the group, this may be by weight, size, duty or other distinguishing factor. The final letter indicates the brake type fitted and will be one of the following:

A Air braked
B Air braked with through vacuum pipe
F Vacuum braked with Accelerated Freight Inshot (AFI)
G Vacuum braked with AFI and through air pipe
H Dual air and vacuum brake with vacuum AFI
O Unfitted vehicle (no automatic brake)
P Unfitted vehicle with through vacuum pipe
Q Unfitted vehicle with through air pipe
R Unfitted vehicle with through air and vacuum pipes
V Vacuum braked
W Vacuum braked with through air pipe
X Dual air and vacuum braked

The three letter TOPS code will be found on the wagon information panel or painted on the vehicle's side, usually in white. When trains are being formed it is most important for this code to be fully understood as trains of mixed types may not be compatible.

When a TOPS office, area freight centre or motive power depot with a terminal requires information about an individual locomotive/vehicle the operator keys in a code for the information required, followed by the locomotive or wagon number including any prefix letters. The required information will then be given either on a video display unit (VDU) or by a hard printer. The TOPS computer is also used for transmitting information of complete train loads from yard to yard, enabling the receiving point to be prepared to accept and deal with the train and its load. Whenever a wagon is moved between yards, whether loaded or empty, all information is given to the master computer. Every wagon to arrive in a yard is checked and its number again given to the computer with details of its next duty, if known, enabling vehicles to be used most economically. Maintenance cycles are also programmed from the TOPS information. Before each train commences a journey the following information is placed on TOPS enabling a train preparation form to be produced by a computer hard printer before departure, and a copy given to the train crew.

1 Wagon numbers (including prefixes) in the same order as they occupy behind the locomotive
2 The loading of all vehicles
3 The individual wagon type (by TOPS code)
4 The length of the train (excluding motive power), progressively totalled from the front wagon by wagon
5 The weight of the train (including motive power) progressively totalled from the front wagon by wagon
6 The brake force (in tonnes)
7 The destination of each wagon (by TOPS figure code)
8 Route directive code

After each train is despatched the TOPS computer is informed and it then gives a 'consist' message to all stopping points for that particular service, giving yard staff a full breakdown of the train due, including motive power, and any repairs or attention needed to vehicles in the formation. Empty wagons are also monitored by TOPS enabling the maximum utilisation of vehicles and keeping empty mileage to an absolute minimum.

The TOPS computer is a large system with vast information recallable in a number of ways. It can produce data about one wagon, a single locomotive's maintenance history, a full list of vehicles at a particular location, or find vehicles fitted with a special modification (ie locomotives fitted with snowploughs), to name but a few. All this information is governed by the code numbers/letters fed into the machine by the operator. However like any computer based system TOPS can only be as accurate as the information fed to it, and therefore cannot be fully relied upon for 100% accuracy at all times.

The departmental TOPS codes used are:

Wagons
Q Former coaching stock vehicles
Y Bogie freight stock
Z Two-axle freight stock

Other types
97 Locomotive stock
92x EMU Stock
93x EMU Stock
968x Former locomotive stock
DX Track machines

Above: Standard wagon plate, with number at the top, weight in the centre, and builder and lot number below. This vehicle No 904152 was introduced in 1960 as a *Lowmac AB* for general freight transportation, however in more recent years the vehicle has been in departmental service as a bogie wheel set carrier on the SR, with its number bearing a 'D' prefix to denote operation under the departmental fleet. *Colin J. Marsden*

Group Q: Departmental coaching stock vehicles

Code	*Description*	*Brake*	*Remarks*
QPA	Canteen coach, staff — dormitory/Staff van	Air	
QPB	Staff van	Air, vacuum piped	
QPO	Staff — dormitory/Staff van	Unfitted	
QPP	Staff — dormitory	Vacuum piped	
QPV	Canteen coach/staff — dormitory/Staff van/ Portakabin coach	Vacuum	
QPW	Staff — dormitory/Staff van	Vacuum, air piped	
QPX	Staff — dormitory/Staff van	Dual	
QQA	Tool van	Air	
QQB	Tool van	Air, vacuum piped	
QQV	Tool van	Vacuum	
QQW	Tool van	Vacuum, air piped	
QQX	Tool van	Dual	
QRA	Stores vehicle	Air	
QRV	Stores vehicle/Denparts/ Battery cell truck	Vacuum	
QRW	Stores vehicle	Vacuum, air piped	
QRX	Stores vehicle/Denparts	Dual	
QSA	Electric car runner/ HST barrier coach	Air	
QSB	Crane runner	Air, vacuum piped	
QSO	Crane runner	Unfitted	
QSP	Crane runner	Vacuum piped	
QSV	Crane runner/Loading ramp/ Train pre-heating vehicle/ Barrier wagon (CS)/Barrier vehicle — non compatible coupling	Vacuum	
QSW	Crane runner/Electric car runner	Vacuum, air piped	
QSX	Crane runner	Dual	
QTA	Brake van	Air	
QTV	Brake van	Vacuum	
QTW	Brake van	Vacuum, air piped	
QVO	Weighing machine van	Unfitted	
QVP	Weighing machine van	Vacuum piped	
QVV	Breakdown equipment/Weed-killing coach/Weighing machine van/cable drum carrier	Vacuum	
QVW	Breakdown equipment	Vacuum, air piped	
QVX	Breakdown equipment	Dual	
QWQ	Leyland inspection vehicle	Air pipe	
QWV	Inspection vehicle (ex DMU)	Vacuum	
QXA	Instruction coach/Generator vehicle/laboratory or test vehicle	Air	
QXB	Laboratory coach	Air, vacuum piped	
QXO	Office coach/Compressor van/ Laboratory coach/Relaying Equipment carrier/Tunnel Platform vehicle	Unfitted	

Code	Description	Brake	Remarks
QXP	Generator Coach/Relaying equipment carrier	Vacuum piped	
QXV	Instruction coach/Film coach/ Inspection saloon/Office coach/Bridge gauge vehicle/ Emergency control unit/ generator vehicle/Laboratory or Test vehicle/Paint spraying vehicle/Pantograph coach/ Rectifier unit/Severn Tunnel Junction Emergency train/ tunnel platform vehicle/Sandite coach/wiring train vehicle, workshop coach	Vacuum	
QXW	Research department driving trailer	Vacuum, air piped	
QXX	Instruction coach/Film coach/ Inspection saloon/Office coach/Generator vehicle/ Laboratory or test coach/ Pantograph coach	Dual	
QYV	Bridge half-deck coach/ Generator coach/Pantograph coach — Electrification/ Wiring coach/Workshop coach — Electrification/Mess or office vehicle/Crane match wagon	Vacuum	

Above: **QPV ADS228.** A large number of former passenger vehicles have been taken into the departmental fleet and are now used as staff transportation or accommodation vehicles. QPV No ADS228 was formerly Ironclad SK No S230. The vehicle is now painted in yellow with black stripes at waist height being used as the crane staff escort vehicle based at Wimbledon. A number of the side doors have been plated over and air vents fitted below window height. *Colin J. Marsden*

Above: **QPV ADS230.** Another former Ironclad which has seen further use in the departmental fleet is S753 an SK, now similarly converted to the above vehicle and is numbered ADS230 but painted in red livery and retains the foot steps on the end. This vehicle is allocated to Hither Green. *Colin J. Marsden*

Below: **QPV ADS231.** Former Ironclad BSK No 3182 was converted during the 1960s into a staff and tool vehicle for use with the breakdown train operating from Wimbledon Park. The car is now numbered ADS231 but seldom used. The former brake portion of the vehicle is now fitted with a large sliding door for the ease of loading equipment. *Colin J. Marsden*

Above: **QPV ADS70130.** Several former Maunsell passenger vehicles still see service in departmental trains. The SR relief breakdown train, kept at Stewarts Lane, has ADS70130 in its formation which was formerly SK No 1023. This vehicle was heavily rebuilt prior to its present use with half of the vehicle stripped of its internal fittings and a tool/equipment area constructed. *Colin J. Marsden*

Above right: **QPV ADS70012.** Now painted in red livery with white numbering and lettering is No ADS70012 formerly Ironclad SK No S746. This vehicle forms part of the Stewarts Lane breakdown train and is solely staff accommodation being internally fitted with seating and cooking facilities. *Colin J. Marsden*

Centre right: **QPV ADE320728.** Operating on the ER are a number of former Gresley vehicles surviving in use as staff and tool vehicles mainly on breakdown trains. The one shown here No ADE 320728, a former third open is painted in yellow livery and is a staff riding van for the Haymarket based breakdown train. *John Tuffs*

Bottom right: **QPV ADB963952.** Formerly a Mk 1 Brake third No Sc 43326, this vehicle is now used as a staff riding coach for the breakdown teams based at March. This vehicle, outshopped late in 1980 from BREL Doncaster, is in yellow livery with black warning strips on the lower part of the body sides and ends. *Michael J. Collins*

Above: **QPV ADS70120.** Another large user of redundant coaching stock vehicles is the Civil Engineering Department (CCE) for use with engineers' trains. ADS70120 formerly SECR BSK No 3557 is now allocated to Woking for use as a Staff and Tool coach on prefabricated assembly trains. It will be noted that some of the side windows have been removed and a large sliding door fitted; the brake section has been retained. *Colin J. Marsden*

Below: **QPV DM395995.** This former LMS CK which now operates on the SR still retains its Departmental Midland (DM) numbering, and works from Woking on civil engineering duties, although in this 1983 view it still sports its home location as Wokingham, a depot long since closed. This vehicle is used to convey staff and small tools to engineering locations and is currently painted dark green. *Colin J. Marsden*

Top: **QPV ADB975424.** The TOPS group Q is not exclusively allocated to bogie vehicles, and a number of short wheelbase freight vans, formerly part of the passenger fleet, come under this classification. QPV No ADB975424 formerly SR PMV No S1107S is now used by the power supply section as a small Staff and Tool van allocated to Horsham. Little conversion work has taken place on this vehicle since its revenue earning days. *Colin J. Marsden*

Above: **QPV DM395799.** The staff escort vehicle for ballast cleaning machine No DR76207 is provided by a former LMS third compartment vehicle No M1451M, now numbered DM395799 and painted in a royal blue livery. All but two of the side access doors have been sealed up and an external lever operate handbrake has been provided. *Colin J. Marsden*

Below: **QPV DM395888.** A further LMS compartment vehicle now forms part of the departmental fleet as a staff accommodation and mess van for the Civil Engineering staff, again painted in blue livery. This vehicle numbered DM395888 is pictured at Derby carriage works, although it is stencilled CCE Department, Crewe. This vehicle is now withdrawn. *Colin J. Marsden*

Above: **QPV DW150359.** This rather unusually liveried vehicle No DW150359 is former Collett BSK No 1595, now being used as a staff escort coach for one of the WR ballast cleaning machines No DR76218 based at Reading. When photographed during 1982 the livery was dark green with yellow doors and ends, and a grey BR logo applied on the bodyside.
Colin J. Marsden

Centre left: **QPV DS70262.** Former Bulleid SO No 1457 is now extensively modified and operates on the SR as an escort and staff vehicle for one of the region's tunnel inspection platforms. The vehicle has had its corridor connection removed, windows altered and an external lever-operated handbrake fitted. The car is seen in green livery at Guildford.
Colin J. Marsden

Below left: **QPV.** A small number of Portakabin wagons have been introduced during the early 1980s mainly for operation on the LMR. The covered staff accommodation is semi-permanently fixed to former Mk I passenger vehicle underframes and is therefore included under the TOPS Group Q. The illustration shows one of the Portakabin units at Water Orton during 1982. *D. Claridge*

Above: **QPV DB975535.** Formerly BSK No S34259 this vehicle is now staff and dormitory accommodation for the soil mechanics section and operates all over the BR network. When not in use it is usually kept at Wimbledon West where the vehicle was photographed during 1982. Again the modification work to this vehicle included the removal of the end corridor and the installation of ventilators under the compartment windows. The brake area remains intact. *Colin J. Marsden*

Below: **QQV ADS70125.** When rebuilding vehicles for departmental usage, considerable work is required for tool vans, normally all original doors and windows are plated over and new openings provided. This was the case with former Maunsell BSK No S3677 which is now the SR relief tool van No ADS70125. Two large sliding doors are provided on each side for equipment access with ventilation grilles at one end where a small generator is situated. The vehicle has yellow livery with black side and end stripes. *Colin J. Marsden*

Above: **QQV ADB975055.** No ADB975055 formerly E70584E is the tool van at Thornaby which was a Thomson bogie brake; again heavy side modification has been made with two large sized sliding doors fitted, although one small window has been kept on each side. The future of some of these older vehicles must now be in jeopardy with large numbers of Mk I vehicles becoming redundant. *Colin J. Marsden*

Below: **QQX ADB975482.** A more recent conversion to a tool van than the previous illustration, is this dual braked vehicle allocated to Gateshead, converted from BSK No E34602 during the 1970s. Considerable external rebuilding has again been undertaken with the fitting of two large sliding doors, one pair of double opening doors and two single opening doors. This vehicle normally operates with the Gateshead crane No ARDC96700. *Colin J. Marsden*

Above: **QQV ADB975121.** To provide storage space for packing and tool equipment for the Bristol Bath Road allocated crane, the sole member of the CCT-Y motor car carrying wagon fleet, No E94100, was converted in the early 1970s for departmental use, with no major modifications needed. Livery during 1983 was bauxite brown this vehicle is now withdrawn. *Colin J. Marsden*

Below: **QRV ADM40263.** For the conveyance of stores from the main BREL works to the smaller servicing depots, a number of former six-wheeled LMS designed X-Fish vans were transferred into the departmental fleet. These vehicles were particularly suitable for stores traffic with two large sliding doors already available on each side, therefore they required very little conversion work before their new use. No ADM40263 is seen in ex-works condition at Doncaster painted in dark green with white lettering and a grey roof. *Colin J. Marsden*

Above: **QRV CDB975832.** A few former GWR designed Siphon G ventilated bogie vehicles are still in traffic within the departmental fleet, mainly acting as stores vehicles. No CDB975832, formerly No W1025W is now devoid of its corridor connections and used for the conveyance of seat trimming materials between BREL Derby and Swindon works. *Colin J. Marsden*

Centre left: **QRV TDB975967.** To operate with the Chipmans weedkilling train BR provide a bogie stores vehicle for the conveyance of the powdered weed control chemical. No TDB975967 formerly SR designed GUV No S4605S has not changed at all since its revenue earning days, with all doors and windows remaining in their original positions. *Patrick Kingston*

Below left: **QRV ADB975158.** A number of vehicles under the QR classification carry the code or lettering *Enparts* this is only a fancy name for a stores van. No ADB975158 illustrated here were formerly No W185W a GWR designed though BR built gangwayed brake (BG), and was photographed at BREL Swindon works in 1982 where it was in use as a seating store. *Michael J. Collins*

Right: **QRV XDB975957.** A small number of former Special Parcels vans (SPVs) have been converted into departmental use, again these are especially suited to further use as large opening doors are available. This vehicle No XDB975957 is in its original condition but repainted for its latest use of conveying stationery stores between the stores at Crewe and the station. *Ian Gould*

Centre right: **QRV ADB977008.** Again little work was necessary to this vehicle after being taken into the departmental fleet as a tool and stores van for the breakdown train unit at Ayr. The vehicle was formerly BG No E81310 and still retained its Inter-City livery and original door and window layout when photographed during 1982. *John Tuffs*

Below: **QRV ADB975197.** To provide additional storage facilities at the Railway Technical Centre, (RTC) Derby, two former BSK vehicles were internally converted for storage purposes. Windows were painted over on the inside and alterations made to the internal layout of ADB975197 formerly BSK No E34375. The guard's compartment still retains its handbrake but this is now only accessible by the external doors. *Colin J. Marsden*

Above and below: **QRX ADB977063, ADB977113.** These two plates illustrate former BR CCT vehicles that have been transferred to the departmental fleet. The upper illustration shows No ADB977063 formerly CCT No M94867 which has received little modification except for the fitting of dual brake equipment and is now used for the conveyance of BRUTE trolleys in the Portsmouth area. The lower plate shows No ADB977113 formerly No M94772 which has had most of the windows painted out, dual brake equipment fitted, and is now used by the CM&EE department to convey stores to and from St Leonards depot. *Both: Colin J. Marsden*

Above: **QSB DB975414.** A surprising type of departmental vehicle to find classified under the former passenger vehicles code are a number of crane and ballast machine match wagons, however as these are built on former passenger vehicle underframes they therefore come into this group. No DB975414 is the match and equipment carrier to WR ballast cleaning machine No DR76218 and is a wooden bodied vehicle assembled on the frame of former BSK No E34499. *Colin J. Marsden*

Below: **QSB DS70277.** To provide match and stores facilities to some of the SR CCE diesel cranes a small number of former EMU underframes were utilised. QSB No DS70277 operating with 15-ton diesel crane No DRT81339 and allocated to Woking, is formerly 6PAN TFK No S12275, now devoid of all its original bodywork. During rebuilding work a through vacuum brake pipe was installed and an external wheel handbrake fitted. *Colin J. Marsden*

Above: **QSA ADB975867.** In recent years a number of special match wagon conversions have taken place and in some cases vehicles with different types of couplings at either end have been introduced, this mainly involved the fitting of *Tightlock* type couplings, as fitted to most modern EMU sets. QSA No ADB975867 introduced in 1982 was rebuilt from RSO No S1006 and is fitted with both high and low level air pipes, waist level EMU control jumpers, *Tightlock* couplings at one end and a buckeye coupling at the other. The equipment on this vehicle is such that conventional EMU and SR locomotives can operate in multiple with Class 455/508 EMUs. QSA No ADB975867 is allocated to Strawberry Hill. *Colin J. Marsden*

Left: **QSA ADB975972.** A number of EMU match wagons were converted during the late 1970s following the introduction of 1972 design high-density stock. The coaches are normally stripped of all existing fittings, external handbrakes fitted and double opening access doors provided for the carriage of stores. One end of the vehicle retains standard drawgear whilst the other has low level *Tightlock* couplings. No ADB975972 was converted from RUO No E1039. When photographed in 1981 this car had suffered end damage.
Colin J. Marsden

Below left: **QSA ADB975999.** FO No E3054 was converted at Derby Litchurch Lane during the early 1980s to act as a match wagon for Mk III sleeping car stock whilst undergoing trials. No external modifications to this vehicle were undertaken, and it is likely to be converted for further departmental use at a later date.
Colin J. Marsden

Above: **QSA ADB975666.** Following the introduction of IC125 trains a fleet of match wagons were converted to be coupled between conventional coupling and IC125 coupling fitted stock. No ADB 975666 formerly Mk II BFK No W14078, painted in rail blue livery with white lettering, stands at Old Oak Common. The internal layout of this vehicle has been retained in its original condition. *Colin J. Marsden*

Below: **QTX TDB975811.** The code QT is given to ex-passenger carrying vehicles now used as brake or guard's accommodation cars. A requirement of all trains is that a position is provided for a guard, and for trains travelling south from BREL Glasgow works, where a guard's van may not be incorporated in the train formation, No TDB975811 was converted. The vehicle was formerly BSK No Sc34724 and retains its guard's van facilities but all other fittings are removed. *Ian Gould*

Above: **QVV DE320995.** The QV TOPS code is given to a number of miscellaneous vehicles including those of BR's Eastern Region weed control train allocated to York. This former Gresley brake third No E16551, now numbered DE320995 is a spray vehicle, the equipment of which can be seen on the nose end. Whilst under conversion from a brake vehicle this car gained end observation windows for the spray operator. DE320995 was photographed at Lowestoft during 1983. *Brian Morrison*

Centre left: **QVB ADB977122.** This vehicle is apparently carrying an incorrect TOPS code, as according to official records it was rebuilt from RB No M1751 to a staff coach for the Toton breakdown crews. It should therefore have been given the code QPB. When photographed at Toton during the summer of 1983 it had grey livery and little internal alteration had taken place. *Colin J. Marsden*

Below left: **QVX ADB975918.** During 1981 BREL Horwich works carried out a number of overhead line maintenance train conversions, rebuilding redundant Mk I passenger vehicles to flat top, or triple drum vehicles. No ADB75918 is built on the frame of former TSO No E4225 and is now equipped to carry three cable drums of overhead power line. *Colin J. Marsden*

Above: **QVX ADB975906.** During conversion of the triple drum vehicles the complete body is removed and a new steel mesh floor provided with waist height hand rail surround. As these vehicles would frequently be required to operate at night, special high powered floodlights were installed. End ladders and a frame mounted handbrake were also provided. No ADB975906 was rebuilt on the frame of SK No E25161. *Colin J. Marsden*

Below: **QXX ADB975681.** To operate with the triple drum vehicles Horwich works also converted a number of Flat Top coaches, these are former Mk I passenger vehicles that have had their roofs lowered and flattened to enable staff to stand on the roofs to attend to overhead wire equipment. Some vehicles are also fitted with pantographs — these are not for power collection but purely for testing purposes. No ADB975681 a pantograph fitted flat top, rebuilt from BSK No W34781 is shown here as part of the Carstairs electrification train. *John Tuffs*

Above: **QXV TDM395280.** Over the years a number of inspection or observation vehicles have been constructed. Normally these are short wheelbase vehicles that have an observation position at each end with a small kitchen and toilet in the centre. The SR now uses one of these vehicles for train crew route training.
No TDM395280 was formerly the DCE inspection vehicle based at Watford. This coach now carries all blue livery and is normally kept at Clapham Junction. *Colin J. Marsden*

Centre left: **QXV TDM45026.** A similar vehicle to the previous plate is No TDM45026 which is still in use as an engineers' inspection saloon based at Walsall. This vehicle has a large and small observation end, each fitted with a central desk with loose chairs down the sides and end; a toilet and small kitchen is provided in the centre of the car together with a guard's compartment. TDM45026 is photographed at BREL Doncaster Works after repairs during 1982. *Colin J. Marsden*

Below left: **QXV DW80975.** The Divisional Engineer on the WR operates two DW8xxxx vehicles, based at Old Oak Common and Bristol.
No DW80975 illustrated here is allocated to Bristol, and is of similar type to those previously described with two observation ends and kitchen etc in the centre. This vehicle is painted in Inter-City livery and has a maximum speed of 90mph (when hauled). *Colin J. Marsden*

Above: **QXV DBM999504.** The LMR operate a fleet of three 999xxx vehicles based at Preston, Newton Heath and Camden, however they can be seen operating all over the network wherever required. These vehicles are of BR design and based on the Mk I features. No DBM999504 illustrated should be numbered TDB999504 and is allocated to Camden for use of the Divisional Manager based in London. The vehicle can either be hauled or propelled and again a large and small observatory are provided.
Patrick Kingston

Below: **QXV DB999509.** Few of the inspection saloon vehicles are fitted with buckeye couplings, however the one allocated for the use of the Divisional Engineer at Newport (Gwent) is so fitted. This vehicle differs from those previously illustrated by having only two front windows, fitted with screen wipers. No DB999509 is seen at BREL Doncaster Works whilst undergoing a refit. *Colin J. Marsden*

Above and below: **QXX DM395223/DW139.** Both the LMR and WR operate track recording coaches to check the alignment and ride of the track. The upper plate shows the LMR vehicle No DM395223 which was converted from an LMS kitchen car No 30005. This vehicle is normally attached to the rear of passenger services and can carry out checking at line speeds. The lower plate is of the WR vehicle No DW139 formerly a brake third No 2360 originally built in 1911. The coach is now equipped with dual brake and heat equipment and mounted on B4 bogies to enable operation with modern rolling stock. When the vehicle detects any irregularity in the track a coloured dye is sprayed on to the track enabling the local staff to deal with the problem. *Patrick Kingston/John Tuffs*

Right, top to bottom:

QXV ADM395838. This interesting former LMS CCT No M35098M is now a mobile generator vehicle belonging to the CM&EE department at Crewe, although photographed at Manchester. The diesel engine and generator set are mounted on the left with a control desk in the centre and a fuel tank on the right. Livery is mid-green with white lettering. *Colin J. Marsden*

QXA ADB975325. To provide an electric supply for ETH fitted rolling stock, and in more recent years IC125 vehicles requiring operation without their correct power cars, a BG No M81448 was converted into a generator van in 1968. The vehicle saw little use until 1980 when, due to the late delivery of IC125 power cars, a train was formed of spare IC125 trailer vehicles with the hitherto redundant generator van providing the higher than standard voltage required on Mk III rolling stock for lighting and heating. The vehicle is able to operate at 100mph and is fitted with air brake equipment only. Between the bogies is a fuel tank and battery box. *Colin J. Marsden*

QXV DW150404. This former Hawksworth BSK No 2220 is now a staff mess and dormitory vehicle on the WR operating with the Neptune track recording train. During the rebuilding of this vehicle the double leaf doors to the brake portion have been sealed up and air ventilators provided under the compartment windows. The livery applied to this coach during 1981 was black with yellow ends and side doors. *Colin J. Marsden*

QXV ADB975375. To provide accommodation for CM&EE staff under training at Wimbledon Park depot, former Bulleid coach No S2526S was converted into a mobile instruction car. The interior of the vehicle was completely rebuilt with a blackboard at one end with rows of seating along the rest of the vehicle. ADB975375 was first numbered in the internal user fleet as No 082232. The livery applied is BR blue with black/yellow wasp ends. *Colin J. Marsden*

Above: **QXV DE321047.** RFO No E3 was rebuilt during the 1970s as a mobile signal power supply generator and is now numbered DE321047 and parked (out of use) in sidings at Barking; livery is pale grey with white lettering. A substantial diesel engine and associated generator set is fixed in the centre of the vehicle in the area of the grilled windows, access being gained by the roller shutters on the ends. *Michael J. Collins*

Left: **QXV ADB977114.** A major problem to receive publicity on a number of occasions over the last few years has been the lack of adhesion when leaves fall on to the running rails during the autumn; they get wet and are crushed by trains which forms a greasy slime causing wheelslip when the next train passes over, both wheel sets and rails are often damaged. To overcome this problem a number of redundant DMU vehicles have been converted to *Sandite* spray cars which carry a substance of sand, gravel and water for distribution on the running rails. These vehicles are not self propelled being hauled/propelled by a diesel locomotive. The laying jets can be seen in front of the wheels of the near bogie in this view of car No ADB977114 formerly DMU Car No M56129 at Cricklewood. *Michael J. Collins*

Below left: **QXV DB975007/008.** This two car set formed of Derby built DMU vehicles Nos 79018/79612 operates all over the BR network as an ultrasonic test set, giving visual and hard readouts of track conditions. The set is fitted with much additional underfloor mounted equipment and is painted in BR blue livery with white lettering and numbering. Normally the set operates over each of the main routes in a strict cycle; when not in use it is usually kept at Old Oak Common. *Colin J. Marsden*

Above: **QXV TDB975994.** A number of former single power cars have been converted into Driver's/Guard's route knowledge training vehicles. The interiors have had their original seating removed and replaced with height staggered seats enabling all those under instruction to have a good view of the line ahead. Few external alterations to these vehicles are usually carried out and their appearance sometimes leaves a little to be desired. No TDB975994 illustrated was formerly No Sc55014.
John Tuffs

Below: One of the most interesting sections of the Departmental fleet are those vehicles operated by the Research Centre based at Derby. A considerable number of redundant vehicles have passed to this section for various requirements, the majority being rebuilt for a specific purpose and often mounted on special bogies, fitted with different couplings or internal test equipment. Some of the vehicles' test purposes are on BR classified listings and therefore only a basic description is given. Two departmental sections operate from Derby, the R&D (Research & Development) unit whose vehicles carry an RDB prefix to their numbers, and the Derby CM&EE whose cars are prefixed by the letters ADB. This plate shows RDB975046 part of the tribometer test train which was converted from BSK No M34249.
Colin J. Marsden

Above: **QXA RDB975000.** This coach converted from RSO No M1003 is now Laboratory 1, being heavily rebuilt for its departmental life as a 125mph vehicle/track interaction instrumentation coach. A lower roof section has been formed at one end to accept a pantograph (this being part of previous trials). Driving controls and an observation position have been fitted at one end and LC1 bogies.
John Tuffs

Below: **QXA RDB975386.** This vehicle now named *Hastings* or Laboratory 4, is rebuilt from a former Hastings line DEMU buffet car No S60750, and is used for tilt and active suspension testing; it is mounted on modified APT style bogies. *Hastings* is authorised to travel at 110mph and is fitted with bar couplings, and special cable ways along the body sides. *Colin J. Marsden*

Bottom: **QXA RDB975636.** This purpose built research vehicle was constructed for APT development work and is formed of welded aluminium. It operates as part of the APT-POP train used to develop tilt, suspension and brakes for the production electric APT fleet. In 1983 the vehicle was used to test electric tilt actuators being developed for further stock. The car usually operates with RDB75634/5 (PC3/4) which are power test cars also for APT development. *Colin J. Marsden*

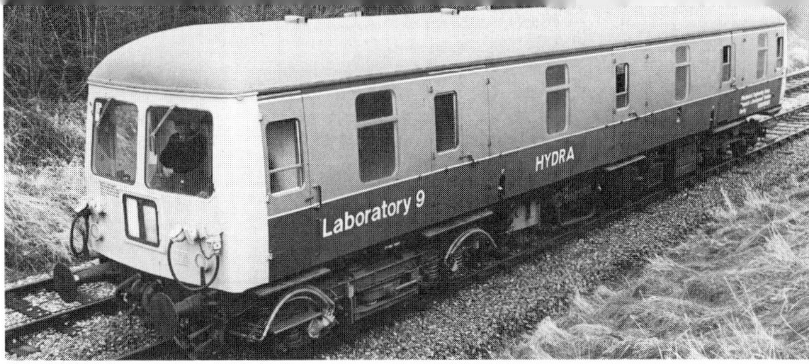

Top: **QXV RDB975385.** Formerly single car No M55997 this vehicle is now named *Hydra* or Laboratory 9, and used for hydraulic transmission and brake testing. The car is operated by the power systems section at Derby and mounted on modified B4 bogies, with a maximum speed of 70mph. The livery is red/blue with yellow warning ends and white lettering. *British Railways*

Above centre: **QXX RDB975076.** BSK No E34500 was converted to the R&D sections 'Auto-Trailer' with full driving controls and all end fittings comparable to EMU/DMU stock, including a four position headcode box. The coach is currently used with the Tribology section for vehicle/ track interaction; it is mounted on B4 bogies and has a maximum speed of 90mph. *Colin J. Marsden*

Above: **QXA ADB975814.** Although this vehicle is operated from Derby Technical Centre it is owned by the DM&EE department and not the Research & Development unit. The car was rebuilt from Class 252 (HST) FO No W41000, and is now Test Car No 10 involved in high speed development operations. Some of the windows have been replaced by grilles where a small internal generator is fitted. *Colin J. Marsden*

Above: **QXX RDB975136.** Converted from BSK No M34505, this car is now Laboratory 12 but only used as a match vehicle within test trains and retains a number of its original facilities. No RDB975136 is fitted with dual brakes, steam heating and has a maximum speed of 100mph. During 1984 the vehicle was operating with the suspension test section.
Colin J. Marsden

Centre left: **QXX RDB975081.** This vehicle resembling a 1963 design SR EMU is in fact a rebuilt BSK No M35313 and is Laboratory 17 formerly named *Hermes*. The coach can provide full driving controls for test trains and is also fitted with a considerable amount of instrumentation equipment; it is mounted on Commonwealth bogies and has a maximum speed of 100mph. It is currently used with the structure gauging train. *Colin J. Marsden*

Below left: **QXX RDB975427.** Rebuilt from Pullman Parlour First with kitchen No M323E, No RDB975427 *Wren* or Laboratory 14, is now part of the test train investigating wheel/rail noise generation and vibration during running which can be measured by special on-board equipment. This car was formerly used in the development of an analytical method for assessing the acoustic design of passenger vehicles. *Wren* is mounted on Commonwealth style bogies fitted with dual brake equipment and authorised to travel at 100mph. During 1984 this vehicle was operating with only one braked bogie.
Colin J. Marsden

Above: **QXV RDM395840.** Laboratory 26 or RDM395840 is rebuilt from an LMS designed SO No M9023M, and is one of the few non-standard BR designed vehicles in operation with the R&D department. The car has been extensively rebuilt for departmental use with a lower roof section able to accept a pantograph with an adjacent observation position. This car is currently used in the development of ultrasonic flaw detection equipment. *Colin J. Marsden*

Below: **QXV ADB975397.** This vehicle is operated by the DM&EE department at Derby and is rebuilt from BSK No M35386 and titled *Test Car 2*. During rebuilding work this vehicle retained much of its original design except for ventilation grilles being positioned in one of the former luggage van doors adjacent to a small generator. This vehicle is currently used for braking tests. *Colin J. Marsden*

Above: **QXX ADB975091.** One of the more frequently observed test vehicles is *Mentor* (Mobile Electronic Network Tester and Observation Recorder). This car was rebuilt from BSK No W34615 in the 1970s and carries the additional indentification of *Test Car 3*. The vehicle is fitted with flat roof sections at each end, both able to accommodate a pantograph, with a central observation position. End observation windows are also provided. Vehicle ends are painted yellow, enabling the car to be formed at the front of a train if required. *Colin J. Marsden*

Below: **QXA ADB975290.** The majority of test vehicles are converted from redundant Mk Is, however *Test Car 6*, introduced in the mid-1970s, was rebuilt from Mk II FK No S13396. Rebuilding work included the fitting of a diesel generator at one end with associated air louvres, and HST style jumpers on the ends. The vehicle is mounted on B4 bogies and has a maximum speed of 100mph. *John Tuffs*

Bottom: **QXV RDB975010.** Derby works built DMU Driving Motor Brake Second No M79900 was taken over by the departmental board after the end of its useful service. The vehicle became the property of the R&D section in 1974 and extensively rebuilt for radio survey work involving UHF and VHF wavebands. The vehicle still retains its driving controls at each end and is now used for radio propogation work, noise measurement, and data transmission tests over radio links. *Colin J. Marsden*

Above: **QXO RDB975311.** Rebuilt from a former SR BY van No S423S is a short wheelbase four-wheeled vehicle now painted in departmental red/blue livery and used as a generator van for site working. Large ventilation grilles have been fitted in the ends and sides but it has no continuous brakes and its maximum speed is 35mph. *Colin J. Marsden*

Below: **QXA RDB975984.** Following the disbanding of the prototype Class 252 High Speed Train the vehicles were distributed around the network, some being rebuilt for further revenue earning service and others being handed over for departmental usage. TRUB No 40000 was renumbered during the early 1980s to RDB975984 and is at Derby RTC waiting possible conversion to a test vehicle. *Colin J. Marsden*

Top: **QXV ADB975968.** After the serious sleeping car fire at Taunton on 6 July 1978 a section was formed at Derby Technical Centre called the Fire Technology Team whose responsibility it was to develop safer stock and equipment. A number of vehicles were taken over by this section and alterations made and eventually they were set afire under controlled conditions to evaluate the results. This car was rebuilt from FK No E13116 and looks rather the worse for wear at Derby in 1981. *Colin J. Marsden*

Above: **QXA ADB975809.** The SR has a small research department based at Strawberry Hill, near Twickenham, where a number of test vehicles are housed, normally associated with EMU traction. Two vehicles were rebuilt in the early 1980s for chopper control tests and named *Romeo* and *Juliet*, one fitted with Brush equipment and the other GEC. No ADB975809 rebuilt from TSO No S3834 is illustrated, and is now fitted with traction motor bogies at each end, shoe gear, high level EMU connections and body mounted resistance frames. *Colin J. Marsden*

Below: **MAG-LEV.** The Research and Development section at Derby has worked on the magnetically levitated rail transport system for a number of years, including the building of a special test track and vehicle. This prototype has now been developed into production cars for the new Birmingham Airport rapid transit system. MAG-LEV is seen 'levitated' at Derby during 1982. *Colin J. Marsden*

Above: **QXA DB999550.** The Mk IIF production run of the 1970s TSO/FO vehicles was extended by one vehicle body, enabling a purpose built track recording vehicle to be constructed. During the build one of the full size side windows was replaced by a half-sized window on both sides and an enormous amount of highly sophisticated test equipment installed including a computer and highly sensitive underframe mounted recording equipment. A floor mounted railbus engine is also provided to give additional power. The car is housed at Derby but its usually on the system covering all routes on a strict timetable basis. In the early 1980s the car, along with its BR staff, was loaned to some European networks. DB999550 is normally attached to the rear of service trains, or formed in special sets for recording purposes. *Colin J. Marsden*

Below: **QYV LDB975066.** Rebuilt from SK No M24140 during the 1970s, this vehicle is now operated by the electrification section, providing staff accommodation and messing facilities on major sites. The car received a major rebuild for its departmental use with corridor connections being removed and a large sliding door fitted at one end for equipment loading. This vehicle is currently located in the East Anglia area whilst the electrification project on the GE Section is underway. *Colin J. Marsden*

Group Y: Departmental bogie freight vehicles

Code	*Name*	*Description*	*Capacity*	*Brake*	*Remarks*
YAA	Brill	Rail/Sleeper	58t	Air	Ex BDA
YAO	Dolphin	Rail/Sleeper	30.5-40.5t	Unfitted	
YBA	Sturgeon	Rail/Sleeper	51t	Air	
YBB	Sturgeon	Rail/Sleeper	51t	Air, vacuum piped	
YBO	Sturgeon	Rail/Sleeper	51t	Unfitted	Some fitted with four bolsters while others have guide angles and timbers
YBP	Sturgeon	Rail/Sleeper	51t	Vacuum piped	Some fitted with guide angles and timbers
YBQ	Sturgeon	Rail/Sleeper	51t	Air piped	
YCA	Halibut	Bogie ballast	52t	Air	
YCO	Pilchard	Ballast/Sleeper (with drop sides)	20.5t	Unfitted	
YCV	Turbot	Ballast/Sleeper	34t	Vacuum	Ex BEV
YDA	Skate	Bogie skip storage wagon	46t	Air	
YEA	Porpoise	Long welded rail carrier	61t	Air	Some with end shute
YGB	Seacow	Ballast hopper (Side/centre chute)	41t	Air, vacuum piped	
YGH	Sealion	Ballast hopper (Side/centre chute)	40.5t	Dual (AFI)	
YGV	Walrus	Ballast hopper (Side/centre chute)	40.5t	Vacuum	
YHA	Whale	Ballast hopper (Side/centre chute)	51t	Air	
YJB	—	Track layer — BREL/Cowans Sheldon/ Plasser	—	Air, vacuum piped	
YJO	—	Track layer — BREL	—	Unfitted	
YJP	—	Track layer — BREL/Cowans Sheldon	—	Vacuum piped	
YJV	—	Track layer — BREL/Thomas Smith/ Taylor Hubbard	—	Vacuum	
YJW	—	Track layer — BREL	—	Vacuum, air piped	
YJX	—	Track layer — Thomas Smith	—	Dual	
YKA	Manta	Long welded rail carrier — 62ft	14t	Air	
YKA	Marlin	Long welded rail carrier — 67ft	14t	Air	
YLA	Mullet	Bogie-rail	58t	Air	Ex BRA

YLO	Gane	Bogie-rail	35.5-40.5t	Unfitted
YLP	Gane	Bogie-rail	40.5t	Vacuum piped
YMA	Salmon	Bogie-rail	51t	Air
YMB	Salmon	Bogie-rail	51t	Air, vacuum piped
YMO	Salmon	Bogie-rail	51t	Unfitted
YMP	Salmon	Bogie-rail	51t	Vacuum piped
YNB	—	Bogie-rail/bolster	57.5t	Air, vacuum piped
YNO	—	Bogie-rail/bolster	28.5-42.5t	Unfitted
YNO	Prawn	Bogie-rail/bolster (no sides)	30.5-51t	Unfitted
YNO	Shrimp	Bogie-rail/bolster (with sides)	42.5t	Unfitted
YNO	Whelk	Boplate	42.5t	Unfitted
YNP	—	Bogie-rail/bolster	30.5-42.5t	Vacuum piped
YNP	Prawn	Bogie-rail/bolster (no sides)	30.5t	Vaccum piped
YNP	Shrimp	Bogie-rail/bolster (with sides)	30.5t	Vacuum piped
YNQ	—	Bogie-rail/bolster	42.5t	Air piped
YNR	—	Bogie-rail/bolster (with sides)	40.5t	Dual piped
YNV	—	Bogie-rail/bolster	30.5—32.5t	Vacuum
YNV	Prawn	Bogie-rail/bolster	30.5t	Vacuum
YOB	—	Travelling bogie crane — Cowans Sheldon/Plasser/Thomas Smith/ Taylor Hubbard	12-45t	Air, vacuum piped
YOP	—	Travelling bogie crane — Coles	10t	Vacuum piped
YOR	—	Travelling bogie crane — Jones/ Thomas Smith	6.5-10t	Dual piped
YOV	—	Travelling bogie crane — Atlas/Warwell	5t	Vacuum
YOW	—	Travelling bogie crane — Jones	6.5t	Vacuum, air piped
YRO	—	Bogie stores & materials — Denparts/ bogie transformer/boiler/machinery	20.5-51t	Unfitted
YRP	—	Bogie stores & materials — Denparts/ bogie transformer/boiler	25.5-51t	Vacuum piped
YRQ	—	Bogie stores & materials — Denparts	10-50t	Air piped
YRR	—	Bogie stores & materials — Denparts/ bogie transformer/boiler	40.5t	Dual piped
YRV	—	Bogie stores & materials — Denparts/ general materials vehicle (ex boplate), bogie transformer/boiler/machinery bogie/ BREL stores	8-51t	Vacuum

Code	Name	Description	Capacity	Brake	Remarks
YSB	—	Crane runner	—	Air, vacuum piped	
YSO	—	Crane runner	—	Unfitted	
YSP	—	Crane runner	—	Vacuum piped	
YSR	—	Crane runner	—	Dual piped	
YSV	—	Crane runner	—	Vacuum	
YSW	—	Shunting truck	—	Vacuum, air piped	
YTV	—	Bogie brake van	—	Vacuum	
YTX	—	Bogie brake van	—	Dual	
YVA	—	Trolley flat/wheel wagon	—	Air	
YVO	—	Concrete saddle arch carrier/girder carrier/trolley flat/cripple carrier/ Breakdown equipment/cable drum carrier	—	Unfitted	
YVP	—	Concrete saddle arch carrier/trolley flat/ cripple carrier/cable drum carrier/wheel wagon	—	Vacuum piped	
YVQ	—	Trolley flat	—	Air piped	
YVR	—	Trolley flat/cripple carrier	—	Dual piped	
YVV	—	Cripple carrier/breakdown equipment/ cable drum carrier	—	Vacuum	
YVW	—	Trolley flat/cripple wagon/wheel wagon	—	Vacuum, air piped	
YXA	—	Test vehicle	—	Air	
YXO	—	Rail mounted auger/relaying equipment carrier/Research department vehicle/ Tunnel platform vehicle/wiring train vehicle/gantry carrier (single line relayer)	—	Unfitted	
YXP	—	Relaying equipment carrier/Research department vehicle/Trenching vehicle/ viaduct inspection vehicle/gantry carrier (single line relaying).		Vacuum piped	
YXR	—	Gantry carrier (single line relayer)	—	Dual piped	
YXV	—	Research department vehicle/Tunnel platform vehicle/Viaduct inspection unit/ Gantry carrier	—	Vacuum	
YXW	—	Tunnel Inspection unit	—	Vacuum air piped	

YYO	—	Electric project — Auger (Rail Mounted)/ Steel carrier/Large plant carrier	—	Unfitted	Ex Weltrol/Bogie bolster/Boplate
YYP	—	Electric project — Auger (Rail Mounted)/ Erection job wagon/Headspan storage/ runner or match/Steel carrier/Concrete mixer/Brake van	—	Vacuum piped	Ex Weltrol/Bogie bolster/Boplate
YYX	—	Electric project — brake van	—	Dual	

YAO DB994070 *Dolphin.*
The 1950/1 fleet of 90 bogie rail/sleeper wagons Nos DB994000-089 were constructed by Head Wrightson as a development of an LNER design with a carrying capacity of 40.5 tons. For loading, the sides of the wagons could be dropped in three sections. No continuous braking was provided but a wheel operated handbrake was available on each bogie. *John Tuffs*

Below: **YBO DB994729 *Sturgeon.*** The BR design variant to the *Dolphin* was the *Sturgeon*, of the same basic appearance but permitted to carry a heavier load of up to 50 tons. One major recognition difference between the two types is the removable ends on the *Sturgeon* and an increase in the number of side doors from three to seven. *Colin J. Marsden*

Above: **YBA DB994502 *Sturgeon.*** The number range for the *Sturgeon* type is DB994090-999 and all were built by Head Wrightson. Some members of the fleet have had their side doors removed for the transport of rails and other CCE equipment. No DB994502 illustrated, is a vehicle operating on the SR and fitted with air brake equipment. *Colin J. Marsden*

Below: **YBO *Sturgeon.*** The *Sturgeon* fleet is usually used for transporting 60ft pre-assembled rail sections or lengths of rail, but they are not suitable for the transportation of continuous welded rails. The bogies on this type are of box construction with the axle boxes built into the side frame.
Colin J. Marsden

Above: **YCV DB978979 *Turbot.*** A more recent introduction into the departmental fleet has been the *Turbot* ballast wagon, with a capacity of 31 tons. These are major rebuilds of BEV *Bogie Bolster E* vehicles. The rebuilding work retained the underframe with a completely new top section giving three drop doors on each side for full length access if required. 1,000 of these are now in service painted in the new departmental colours of grey and yellow. *Colin J. Marsden*

Below: **YGB DB980024 *Seacow.*** For transporting ballast to permanent way sites a number of bogie ballast wagons are in service with side and/or centre chutes. Three types of vehicle are classified under YG — the *Seacow, Sealion* and *Walrus*. This plate shows the *Seacow* type which has a 40 tons capacity and is a development of 1947 designed *Walrus* fitted with air brakes and vacuum pipe. No DB980024 is shown in grey/yellow livery. *Colin J. Marsden*

Above: **YGH DB982680 *Sealion.*** From 1971 a subsequent build of bogie hopper vehicles was introduced of identical design to the *Seacow* but fitted with dual brake equipment with AFI (Accelerated Freight Inshot) being classified as *Sealion*. The three wheels on each end of the vehicles on the verandah operate the left, centre and right ballast chutes. Under the ballast wheels at the near end in the illustration are the vacuum brake cylinders, whilst at the far end the air brake distributors are situated. *Colin J. Marsden*

Below: **YGH DB982682 *Sealion.*** This *Sealion* hopper is painted in olive green livery with white lettering. The number range for the *Seacow* fleet is DB982540-567, whilst the *Sealion* fleet are numbered DB982440-539/568-927. The *Sealion* vehicles are allocated to all regions and can be seen operating over the majority of the network, however the *Seacow* fleet are allocated only to the SR. *Patrick Kingston*

Above: **YJB DRP78211.** Type TJC60 track lifting crane. Fourteen of these cranes were built by Plasser & Theurer and are currently in operation at various locations throughout BR, operated by the Chief Civil Engineering Department (CCE) being used exclusively for permanent way operations. Each crane is fitted with twin jibs each capable of lifting 7ton independently or 12ton if used together. The vehicles weigh 84 tons and are 15.6m in length. *Colin J. Marsden*

Below: **YJB.** TJC60 crane in operation. When involved in track relaying the TJC60 crane would be placed on the adjacent track to that being re-laid, spanning the gap between the two lines with its telescopic jib. This type of crane is fitted with air brakes and a through vacuum pipe. *Barry J. Nicolle*

Above: **YJB DRC78230.** Of a similar type to the Plasser & Theurer TJC60 is this Cowans Sheldon designed and built twin jib unit which is able to offer the same lifting capability as the TJC60. A central operating position is provided giving good all round visibility. The livery is black underframe with yellow above. *John Tuffs*

Below: **YJW DRB78113.** A small number of BREL built track laying cranes are in operation mainly on the LMR. These vehicles are built on bogie well underframes and conform to the same basic type as those shown previously with twin jibs and able to lift around 12 tons with a central operating position. Although this vehicle is classified as YJW when photographed at Horwich in 1981 no air pipes could be seen. *Colin J. Marsden*

Above: **YKA DB975510 *Marlin*.** During the late 1960s and early 1970s 34 redundant SR EMU express cars gave their underframes to departmental use as long welded rail carriers and titled *Marlin* or *Manta*. The coach bodies were completely removed and the underframe rebuilt with cross bolsters and an external underframe mounted handbrake. All cars are air braked and retain EMU style buffers. Numbering is in the DB975500-33 range. No DB975510 illustrated here was formerly SR car No S10032 which has an unloading chute at the near end. *Colin J. Marsden*

Below: **YLO DW40089 *Gane*.** Two types of *Gane* bolster wagon are in service, those designed and built by the GWR, and those subsequently constructed under BR control. The vehicle shown is of GWR origin, it has a 40 tons carrying capacity, is 62ft in length and used to convey various civil engineering loads. Cross bolsters are usually carried by this wagon type with side support stanchions frame mounted. No DW40089 is seen at Carnforth in 1983. *Colin J. Marsden*

Above: **YMP DB996918 *Salmon.*** A considerable number of 51ton bogie wagons classified as *Salmon* are in traffic. These are basically a development of the *Gane* type and are a bogie flat used for the transportation of civil engineering loads. Wagons were constructed by either Head Wrightson or G. R. Turner Ltd and numbered in the DB996xxx range. The vehicle shown here No DB996918 is carrying a bridge side section at Oxford. *Michael J. Collins*

Below: **YMP DB996915 *Salmon.*** Two different bogie types exist within the *Salmon* fleet; early vehicles constructed by Head Wrightson are carried on short bogies whereas later built GR Turner vehicles have a lengthened plate type bogie. Carrying some 60ft rail sections No DB996915 with long bogies is seen at Willesden in 1982. Although this vehicle has a vacuum pipe it is not brake fitted and is only a through connection. *Colin J. Marsden*

Bottom: **YMA DB996406 *Salmon.*** A number of *Salmon* vehicles have been converted to air brake operation and classified YMA. This was primarily done for long welded rail operation but their use is now more widespread; often AB fitted wagons are painted yellow. No DB996406 illustrated is part of the SR welded rail stock. *Colin J. Marsden*

Above: **YNV DB923942.** A large number of ex-revenue earning bogie flat vehicles have passed into departmental usage operating with various sections. The YNV illustrated is working for the CM&EE department ferrying B4 bogies between depots. This vehicle was formerly a bogie bolster E (BEV) and has seen no external change for its departmental application except a D prefix to the number. *Colin J. Marsden*

Below: **YNP ADB946035.** Many different types of flat bolster style vehicles come under the YN classification with carrying capacities ranging from 25 to 51 tons. This vehicle, again from the ex-revenue fleet, is converted from a Bogie Bolster C (BCP) and now operated by the CM&EE department for bogie transportation. Through vacuum pipes are fitted. *Colin J. Marsden*

Bottom: **YNP ADS57838.** Not only former BR designed vehicles have been transferred to the departmental fleets and this ex SR *Borail* is also operated by the CM&EE for bogie transport. When vehicles are transferred for this kind of use it is unlikely that any modification work would be undertaken. No ADS57838 is seen at Guildford in 1983. *Colin J. Marsden*

Above: **YNP DB997648 *Shrimp*.** This one off vehicle now classified as *Shrimp* is a 30ton bogie bolster with sides, built in 1953 by BR at Lancing to lot No 2947. When constructed the vehicle was intended for S&T department use, but in recent years it has been operated by the CM&EE power supply section based at Horsham. An interesting feature of this vehicle is that the sides can be removed completely and the wagon operated as a flat. *Colin J. Marsden*

Below: **YNB DC920000.** A large 92ton capacity bogie steel flat was constructed by BREL Ashford in 1976 numbered 920000 in the AB numbering series and classified BLA. In 1982 the vehicle was transferred to Departmental stock for the Civil Engineers' use and classified YNB and given the prefix DC to its number. The vehicle has steel lattice floor and removable side stanchions. No DC920000 is seen at Woking in mid-1982. *Colin J. Marsden*

Right: **YOB DRP81511.** Plasser & Theurer type GPC 72 telescopic jib crane, front end detail (jib end): 1. Headlights (used for night operation), 2. Emergency lifting lugs, 3. Through vacuum pipe, 4. Main reservoir brake pipe, 5. Air brake pipe, 6. Coupling hook and screw shackle, 7. Tail lamp brackets. *Colin J. Marsden*

Below: **YOB DRP81521.** During 1984 there were 30 Plasser & Theurer GPC type cranes operating on BR, allocated to all regions and used within the CCE Department for permanent way operations. Each vehicle has a three section telescopic jib which is capable of lifting a maximum load of 12 tons. The vehicles weigh 80 tons when in fully operational condition. DRP81521 is allocated to the SR at Woking. *Colin J. Marsden*

Above: **YOB DRP81511.** As these vehicles are often required to operate at night or in arduous conditions high powered lights on both ends and sides are provided to meet safety requirements. A siren warning system is also available to warn staff that a train is approaching. Livery is black underframe with yellow above solebar height. *Colin J. Marsden*

Below: **YOB DRP81511.** The GPC 72 style machine is powered by a 185kW diesel engine which provides power for traction when operating on site, as well as power for its crane and hoist equipment. All vehicles are mounted on Gloucester style cast bogies. *Colin J. Marsden*

Right: **YOP.** All CCE and CS&TE department bogie cranes are classified as YO, these normally only have a small lifting capacity compared to some of the larger breakdown cranes operating with the CM&EE. The 10ton lift crane illustrated is of the Coles type, operating with the CCE department on the ER. These vehicles are normally painted in yellow livery with black underframes.
Colin J. Marsden

Below: **YOR DRT81339.** A number of Taylor Hubbard 15ton capacity cranes are in operation with the CCE department for permanent way relaying duties. This vehicle No DRT81339 with a tare weight of over 60 tons is based at Woking and powered by a small diesel engine, visible in the rear of the body. When photographed during 1982 this crane was in grey livery and looked rather the worst for wear being a likely candidate for early retirement.
Colin J. Marsden

Above: **YOV ARDW98500.** Operating from the SR power supply section at Horsham is an Atlas/ Warwell crane unit with a lifting capacity of only 5 tons. The Atlas crane jib and controls are mounted on a redundant *Warwell* wagon which has received major modifications to accept controls for the jib, storage boxes and a small stowage area, the vehicle retains it vacuum brakes. This crane is normally used to lift small electric transformers to and from lineside power houses. *Colin J. Marsden*

Below: **YRV ADB900925.** This vehicle, a former *Weltrol WP*, was introduced in 1960 as part of lot No 3192, built by BR Swindon works it is now a locomotive bogie transporter between BREL Crewe and Doncaster works. No major modifications have been carried out for its present use except temporary wooden supports laid in the well to prevent the bogie moving whilst in transport. *Colin J. Marsden*

Above: **YRQ ADB907222.** Another vehicle now utilised to carry parts to and from BREL workshops is this former *Flat ET* No B907222 introduced in 1958. This vehicle has a through air brake pipe but is not brake fitted itself and is mainly used to convey bogies or single wheelsets. No ADB907222 is photographed at BREL Doncaster during 1982. *Colin J. Marsden*

Below: **YSP DW107103.** A number of ex-company built vehicles survive as match wagons for small departmental cranes. WR allocated crane No 81549 has match wagon No DW107103 which is a GWR designed and built wagon, and when photographed in 1982 it still sported a GWR Maker's plate. This vehicle is typical of the GWR match wagon design with pannier equipment boxes each side of the jib position. *Colin J. Marsden*

Above and below: **YTV ADS56283. YTX ADS56286.** This design of bogie brake van dates back to the 1920s when some redundant ac electric stock underframes were utilised as a basis for long wheelbased vehicles designed for express freight operation. The SR built further 'new' bogie brake vans during 1936 and the vehicles illustrated are from this build. The guard's accommodation in these vehicles is no larger but the extended veranda takes up the additional room. They are now operated by the rolling stock engineers (RSE) section and painted in olive green. The upper illustration shows No ADS56283 a vacuum braked example, whilst the lower plate shows No ADS56286 which has been fitted in recent years with dual brake equipment. *Colin J. Marsden/John Tuffs*

Above: **YVP ADB901015.** The TOPS code YV covers a number of different vehicle types, all based on flat or well vehicles. YVP No ADB901015 is one of three *Weltrol WGG* vehicles built in 1960 at Swindon works. It now operates as an equipment carrier for the East Wimbledon breakdown crane, transporting bogies to the scene of any major derailment. The livery is red with white numbering and lettering. *Colin J. Marsden*

Below: **YVP ADB901453.** Another well wagon operated by the CM&EE department with breakdown trains is No ADB901453 formerly the sole *Protrol EF* built by BR Ashford in 1963 for the transportation of propellers. It is now painted yellow with black lettering and allocated to Stewarts Lane depot, where it was photographed during 1982. *Colin J. Marsden*

Above: **YVP ADB900911.** This *Weltrol WZ* was introduced in 1958 as the only vehicle of lot No 2977 constructed at Swindon works. The large vehicle with a carrying capacity of 25 tons is now operated by the RSE section and used to transport bogies between depots. Stencilled on the side of this vehicle is the instruction that if a load is placed centrally in the well the maximum load must be halved.
Colin J. Marsden

Below: **YVP DB900417.** *Flatrol MHH* No 900417 was introduced in 1956 for the revenue earning fleet and used to carry general merchandise until the 1970s when it was transferred to the departmental fleet for use by the CCE department, and when photographed at Carnforth in 1983 had been used to bring a tracked 'Hymak' bulldozer to an engineering site. This vehicle is fitted with a through vacuum pipe and handbrakes at both ends, in common with the previously illustrated YVP types.
Colin J. Marsden

Above: **YVP ADB900810.** No ADB900810 is a *Weltrol MC* introduced in 1959 as part of lot No 3272, constructed at the L&Y Wagon Co and having a carrying capacity of 20 tons. Again this vehicle is operated by the CM&EE department on the Eastern Region and is seen at BREL Doncaster Works loaded with a coach bogie. On the side a stencilled notice states 'Return to Carr loco Doncaster'.
Colin J. Marsden

Below: **YVR DB901744.** This very long well wagon of the *Trestrol EO* type was built by BR Derby works in 1960 to carry 40ton loads and was one of 50 similar wagons. The vehicle has no continuous brake but is fitted with both air/vacuum through pipes, enabling it to be formed in almost any train. It is currently in traffic with the CM&EE department but does not carry the 'A' prefix to the number when photographed in 1983 at Stewarts Lane. *Colin J. Marsden*

Above and below: **YVP ADB976022. ADB976019.** 1972 saw the introduction of 12 long wheelbased cable drum wagons, capable of carrying three drums each, the vehicles were converted at Stewarts Lane Carriage & Wagon works from *Weltrol* wagons. Fleet numbers ADB976016-027 are operated by the CM&EE power supply section based at Horsham and can be seen throughout the SR. Livery is green with white lettering and grab rails. The upper view shows No ADB976022 with one drum at Horsham while the lower plate illustrates No ADB976019 in ex-works condition at Guildford.
Both: Colin J. Marsden

Above and below: **YVP ADB900804/805.** Some more recent triple drum conversions than those shown on the previous page are former *Weltrol MC* vehicles which now operate with the CM&EE power supply section. These are similarly converted to the previously described type, except they are of a shorter wheelbase. Three drums are again carried with the largest in the middle. Both vehicles illustrated are at Guildford, where a number of these wagons operate for the power supply section.
Both: Colin J. Marsden

Above: **YVP ADB909012.** No ADB909012 now operated by the CM&EE department as an equipment carrier is often used for transporting locomotive or coach bogies. It was formerly a *Rectank EC* and one of 21 vehicles built by Butterley Engineering in 1960 under lot No 3165, and has a carrying capacity of 38 tons. This wagon is mounted on Gloucester style bogies and fitted with a through vacuum pipe. *Colin J. Marsden*

Below: **YXP DM748349 TU37.** This vehicle is now a viaduct inspection unit and has seen an eventful life. The well wagon that the unit is mounted upon was built prior to World War 2 as a *Warwell* owned by the War Department; after hostilities ceased it became the property of the LMS who registered it as M748349. After some years the vehicle was rebuilt into a 30ton *Borail* by the addition of bolsters and a floor over the well section. During the 1970s when a viaduct inspection unit was planned this vehicle was again available as surplus and was chosen to be rebuilt as TU37 the BRB viaduct and bridge inspection vehicle. The wagon is now fitted with a steel floor and inspection platform which can be raised for bridge observation, or lowered over the side of a structure for underside inspection. The unit is painted yellow and operates nationwide over the BR network. It is seen here in the sidings at Surbiton in 1982. *Colin J. Marsden*

Above and below: **YXW DM721221.** A number of Tunnel Inspection Gantry vehicles operate throughout BR. No DM721221 on the SR has a steel frame and wooden decked platform mounted on a former LMS *Warwell* of 1942 Wolverton vintage. The platform is swung upright when the vehicle is in its inspection position. Small tool and equipment boxes are kept at one end and a gas storage compartment at the other. The vehicle is painted in departmental yellow livery. *Both: Colin J. Marsden*

Above: **YXW DB901202.** One of the largest inspection vehicles in operation is this roof and structure observation vehicle, built on to former 12-wheel *Weltrol EJC* No DB901202 1965 Lancing vintage, with a carrying capacity of 55 tons. For its latest use the vehicle has been completely rebuilt and carries an *Armfield* hydraulic platform which can either operate in a direct vertical manner or be swung outwards and upwards to ease access. Large control cabins have been positioned at each end. The vehicle operates throughout the BR network and is seen at Waterloo in January 1981 soon after completion and being inspected by various CCE staff. *Colin J. Marsden*

Below: **YXO RDB901601 *Cyclops.*** No this is not something from Doctor Who!! It is Derby Research vehicle *Cyclops* used by the track development unit for track measuring. The vehicle has an accommodation cabin built on former 12-wheel *Trestrol EC* No B901601 built in 1950 by Tees Side SBE. It now has a maximum speed of 35mph and has no continuous brake, only a handbrake. This RTC vehicle is unusual in being housed when not in use at York. Photographed at Knottingley in 1982. *Michael J. Collins*

Above: **YXA RDB602150** ***Decapod.*** Another vehicle that looks as if it has come from another planet is Laboratory No 25 *Decapod*, from the Research Centre at Derby. This interesting vehicle is formed on a former freightliner bolster with a coach style laboratory building assembled in the centre, heavy weight blocks are positioned at both ends. It is used by the vehicle track interaction group and is designated as a lateral track loading vehicle. An additional retractable wheelset is provided under the centre of the car for recording purposes. *British Railways*

Below: **YXA RDB998546** ***Convex.*** This experimental freightliner container flat was built in 1981 by Standard Wagon Co for the Research & Development Section, in an attempt to design a flat with increased payload capacity, reduced mass, increased fatigue life and lower building costs. The vehicle is 60ft in length and fitted with standard twistlock fasteners to accept conventional containers. It is currently undergoing evaluation running in a freightliner formation. *Colin J. Marsden*

Group Z: Departmental stock two-axle freight vehicles

Code	*Name*	*Description*	*Capacity*	*Brake*	*Remarks*
ZAA	Pike	Ballast — sleeper (drop side)	31t	Air	Ex SPA
ZAO	—	Ballast — sleeper (drop side)	13-13.2t	Unfitted	Ex Medium
ZAP	—	Ballast — sleeper (drop side)	13-13.2t	Vacuum piped	Ex Medium
ZAV	—	Ballast — sleeper (drop side)	13-13.2t	Vacuum	Ex Medium
ZAV	Cod	Ballast — sleeper (drop side)	22.5t	Vacuum	Ex Batton
ZBA	Carp	Ballast side and end door	24t or 15.3cu m	Air	
ZBO	Grampus	Ballast side and end door	20.5t or 11.6cu m	Unfitted	
ZBO	Lamprey	Ballast side and end door	20.5t or 13.9cu m	Unfitted	
ZBP	Grampus	Ballast side and end door	20.5t or 11.6cu m	Vacuum piped	
ZBP	Lamprey	Ballast side and end door	20.5t or 13.9 cu m	Vacuum piped	
ZBQ	Grampus	Ballast side and end door	20.5t or 11.6cu m	Air piped	
ZBR	Grampus	Ballast side and end door	20.5t or 11.6cu m	Dual piped	
ZBV	Grampus	Ballast side and end door	20.5t or 11.6cu m	Vacuum	
ZBV	Lamprey	Ballast side and end door	20.5t or 13.9cu m	Vacuum	
ZBW	Grampus	Ballast side and end door	20.5t or 11.6cu m	Vacuum, air piped	
ZCA	Sole	Ballast — sleeper (drop side)	12t	Air	
ZCO	Haddock	Ballast — sleeper (drop side)	12t	Unfitted	
ZCO	Minnow	Ballast — sleeper (fixed sides)	14t	Unfitted	
ZCO	Ling	Ballast — sleeper (drop sides)	14t	Unfitted	
ZCO	Sole	Ballast — sleeper (drop sides)	12t	Unfitted	
ZCO	Starfish	Ballast — sleeper (drop sides)	10t	Unfitted	
ZCO	Tunney	Ballast — sleeper (drop sides)	20.5t	Unfitted	
ZCV	Tunney	Ballast — sleeper (drop sides)	20.5t	Vacuum	
ZCV	Plaice	Ballast — sleeper (drop sides)	22.5t	Vacuum	Ex-plate
ZDA	Bass	General material carrier	31.5t	Air	Ex-SAA/OBA
ZDO	—	General material carrier	10-25t	Unfitted	Ex-Medium/sand/van/low conflat/ plate/tube/mineral/coal hopper
ZDP	—	General material carrier	13-22t	Vacuum piped	Ex-Medium/Gunpowder/low/conflat/ plate/tube
ZDV	Dace	General material carrier	20.5t	Vacuum	Ex-OUV/KLV

ZDV	—	General material carrier	13-33t	Vacuum	Ex-Medium/sand/van/low/conflat/plate/pipe/tube/anhydrite/ironstone/shochood
ZDW	—	General material carrier	12t	Vacuum, air piped	Ex-pipe
ZDX	—	General material carrier	22.5t	Dual	Ex-tube
ZEO	—	Ballast hopper (centre chute)	21.5t	Unfitted	Ex-Ironstone hopper
ZEV	Catfish	Ballast hopper (centre chute)	19.5t	Vacuum	
ZFP	Trout	Ballast hopper (centre and side chute)	25.5t	Vacuum piped	
ZFV	Dogfish	Ballast hopper (centre and side chute)	24.5t	Vacuum	
ZGB	—	General material carrier	13t	Air, vacuum piped	Ex-High
ZGO	—	General material carrier	10-13.2t	Unfitted	Ex-High, some vehicles have drop sides
ZGP	—	General material carrier	13-13.2t	Vacuum piped	Ex-High
ZGV	—	General material carrier	10-13t	Vacuum	Ex-High, some vehicles have drop sides
ZGW	—	General material carrier	21.5t	Vacuum, air piped	Ex-Runner
ZHO	—	CCE spoil	16-27t	Unfitted	Ex Mineral/Tippler
ZHV	—	CCE spoil	16.5t	vacuum	Ex-16.5t Min
ZIB	—	Crane (breakdown) Cowans Sheldon	36-75t	Air, vacuum pipe	
ZIP	—	Crane (breakdown) Cowans Sheldon/Ransome Rapier/Cravens	36-50t	Vacuum piped	
ZIV	—	Crane (breakdown) Cowans Sheldon/Ransome Rapier	45-75t	Vacuum	
ZJV	Mermaid	Ballast (side tipping)	14t	Vacuum	
ZKO	—	Ballast wagon	26.5-27.5t	Unfitted	Ex-Tippler
ZLV	Herring	Ballast hopper (centre chute)	20.5t	Vacuum	
ZMV	Mackerel	Ballast hopper (centre chute)	17.5t	Vacuum	
ZNO	—	CCE steel carrier	21.5t-22t	Unfitted	
ZNP	—	CCE steel carrier	20.5-22t	Vacuum piped	
ZNV	—	CCE steel carrier	13-22t	Vacuum	
ZOB	—	Travelling crane — fixed axle — Cowans Sheldon/Taylor Hubbard	12t	Air, vacuum piped	
ZOO	—	Travelling crane — fixed axle — Coles/Geismar/Russell/Thomas Smith/Taylor Hubbard	5-10t	Unfitted	

Code	*Name*	*Description*	*Capacity*	*Brake*	*Remarks*
ZOP	—	Travelling crane — fixed axle — Booth/Cowans Sheldon/Coles/Geismar/Thomas Smith/Taylor Hubbard	3-15t	Vacuum piped	
ZOR	—	Travelling crane — fixed axle — Booth/Thomas Smith/Taylor Hubbard	10-12t	Dual piped	
ZOV	—	Travelling crane — fixed axle — Cowans Sheldon	30t	Vacuum	
ZPO	—	Staff/Dormitory vehicle	—	Unfitted	
ZPP	—	Staff/Dormitory vehicle	—	Vacuum piped	
ZPV	—	Staff/Dormitory vehicle	—	Vacuum	
ZQO	—	Tool van	7-12t	Unfitted	
ZQP	—	Tool van	12t	Vacuum piped	
ZQV	—	Tool van	10-20.5t	Vacuum	
ZRB	—	General stores vehicle	10-11.5t	Air, vacuum piped	
ZRF	—	Tank, fuel oil/diesel	22.5t	Vacuum (AFI)	
ZRO	—	Stores vehicle for concrete products/general stores. Tank for: Creosote/fuel oil/diesel/lubricating oil/oil storage/sludge/tar/waste oil/water	7-25.5t	Unfitted	
ZRP	—	Stores vehicle for general stores/transformers. Tank for: fuel oil/waste oil/water	7-20.5t	Vacuum piped	
ZRQ	—	Tank for fuel oil/diesel oil/waste oil	20.5t	Air piped	
ZRR	—	Stores vehicle — transformer. Tank for: fuel oil/diesel oil/water	20.5t	Dual piped	
ZRV	—	Stores vehicle — batteries/cells/concrete products/Denparts/gas cylinders/general stores/transformers/BREL Export Stores/Tank vehicles — Bituminous Emulsion/Creosote/fuel oil/diesel oil/lubricating oil/sludge/waste oil/water/weedkiller/anti-freeze/de-icing fluid	6-25.5t	Vacuum	
ZRW	—	Stores vehicle — General stores/transformer. Tank vehicles — water	6-25.5t	Vacuum, air piped	
ZSA	—	Barrier vehicle/runner for snow blower	—	Air	

ZSB	—	Crane runner	—	Air, vacuum piped	
ZSO	—	Crane runner/electric car runner/shunting truck	—	Unfitted	
ZSP	—	Crane runner/shunting truck/train pre-heating unit	—	Vacuum piped	
ZSQ	—	Electric car runner/shunting truck	—	Air piped	
ZSR	—	Crane runner/shunting truck	—	Dual piped	
ZSV	—	Crane runner/electric car runner/loading ramp/mobile walk-way platform/shunting truck/train pre-heating vehicle/Barrier (conflat) for long welded rail train	—	Vacuum	
ZSW	—	Crane runner/electric car runner/mobile walk-way platform/shunting truck	—	Vacuum, air piped	
ZSX	—	Shunting truck	—	Dual	
ZTO	—	Brake van	—	Unfitted	
ZTP	—	Brake van	—	Vacuum piped	
ZTQ	—	Brake van	—	Air piped	
ZTR	—	Brake van	—	Dual piped	
ZTV	—	Brake van	—	Vacuum	
ZUP	Shark	Brake van with plough	—	Vacuum piped	
ZUV	Shark	Brake van with plough	—	Vacuum	
ZUW	Shark	Brake van with plough	—	Vacuum, air piped	
ZVA	—	Wheel wagon	20-25t	Air	Ex-OBA
ZVO	—	Trolley flat/steel carrier/ cripple carrier/ signal equipment carrier/breakdown equipment carrier/weights truck/cable drum carrier/wheel wagon	13-25t	Unfitted	
ZVP	—	Trolley flat/cripple carrier/breakdown equipment carrier/weights truck/cable drum carrier/wheel wagon	14-25t	Vacuum piped	
ZVQ	—	Trolley flat	20.5-21t	Air piped	
ZVR	—	Trolley flat/cripple carrier	20-25t	Dual piped	
ZVV	—	Trolley flat/steel carrier/cripple carrier/ signal material carrier/breakdown material carrier/weedkilling vehicle/weighing machine wagon/weights truck/cable drum carrier/wheel wagon/wheel wagon with crane	11-25.5t	Vacuum	

Code	*Name*	*Description*	*Capacity*	*Brake*	*Remarks*
ZVW	—	Trolley flat/cripple carrier/breakdown equipment carrier/weights truck	12-25t	Vacuum, air piped	
ZVX	—	CCE Equipment carrier/bogie wheel carrier	20t	Dual	
ZWA	—	Self propelled snow blower	—	Air	
ZWB	—	Ballast cleaner, self propelled	—	Air, vacuum piped	
ZWP	—	Ballast cleaner, self propelled	—	Vacuum piped	
ZWV	—	Ballast cleaner, self propelled	—	Vacuum	
ZXA	—	Generator coach/rail crane winch wagon/ roller wagon for rail crane winch/ Research department vehicle	—	Air	
ZXB	—	Tunnel platform vehicle/Research department vehicle	—	Air, vacuum piped	
ZXO	—	Bridge gauging vehicle/generator vehicle/ paint spraying vehicle/rail crane winch wagon/relaying equipment carrier/roller wagon for rail winch wagon/trenching vehicle/tunnel platform vehicle/wiring train vehicle/Research department vehicle/ viaduct inspection ancillary vehicle/brake van for viaduct inspection train/brake van working with crane	—	Unfitted	
ZXP	—	Ballast cleaning machine equipment/rail crane winch wagon/relaying equipment carrier/trenching vehicle/tunnel platform vehicle/wiring train vehicle/refuse skip carrier/Research department vehicle	—	Vacuum piped	
ZXQ	—	Bridge gauge vehicle/relaying equipment carrier	—	Air piped	
ZXR	—	Relaying equipment carrier/brake van working with cranes or ballast cleaners	—	Dual piped	
ZXV	—	Ballast cleaning machine equipment/ bridge gauge vehicle/electric charging plant van/emergency control unit vehicle/ fire engine vehicle/generator vehicle/	—	Vacuum	

		paint spraying vehicle/rail crane winch wagon/relaying equipment carrier/roller wagon for rail crane winch wagon/tunnel platform vehicle/wiring train vehicle/ Research department vehicle/viaduct inspection ancillary vehicle/brake vans working with tunnel inspection unit/brake vans working with cranes		
ZXW	—	Relaying equipment carrier/viaduct or overhead inspection vehicle/brake van working with crane	—	Vacuum, air piped
ZXX	—	Relaying equipment carrier/Research department vehicle	—	Dual
ZYO	—	Electrification project — concreting auxillary/generator vehicle/grouting wagon/steel carrier-crane match wagon	—	Unfitted
ZYP	—	Electrification project — cable drum carrier/concreting auxiliary/excavation grab/grouting wagon/steel carrier/water tank/brake van/match wagon to crane	—	Vacuum piped
ZYR	—	Electrification project — lowmac/single bolster/match wagon to crane	—	Dual piped
ZYV	—	Electrification project — cement van/ concrete track mixer/concreting auxiliary/ excavation grab/generator carrier/goods van (general)/tunnel platform vehicle/ tunnel trestle	—	Vacuum
ZYV	—	Travelling crane — brake van/match wagon	—	Vacuum
ZYW	—	Travelling crane — brake van/match wagon	—	Vacuum, air piped
ZYX	—	Electrification project — cable drum carrier/generator carrier/tunnel platform vehicle/staff van	—	Dual
ZZB	—	Mobile power/load bank (former Class 84 loco)/crane — travelling (diesel)	—	Air, vacuum piped

Code	Name	Description	Capacity	Brake	Remarks
ZZO	—	Generator — static for train pre-heating (ex-locomotive)/snowplough tender/ track relaying crane/crane — travelling — diesel/crane — travelling — manual/ crane — travelling — steam	—	Unfitted	
ZZP	—	Generator — static for train pre-heating (ex-locomotive)/independent snowplough/steam breakdown crane/ track relaying crane/crane — travelling — diesel/crane — travelling — manual	—	Vacuum piped	
ZZR	—	Crane — travelling — diesel	—	Dual piped	
ZZV	—	Independent snowplough/snowplough tender/crane — breakdown — diesel/ crane — breakdown — steam/track relaying crane/crane — travelling — diesel/Class 24 service locomotive	—	Vacuum	

ZAA DC460049 *Pike*. A recent addition to the four-wheeled departmental ballast fleet has been 320 ZAA *Pike* wagons converted from ex-revenue fleet SPAs. The only alteration is the addition of the DC prefix to the number and the name *Pike* stencilled on the side. In the majority of cases the *Railfreight* logo has been removed. These vehicles are intended as general transport carriers. ZAA No DC460049 was photographed at Newton Abbot. *Colin J. Marsden*

Left, top to bottom: **ZBO DB986348 *Grampus.*** The BR standard drop side ballast wagon adopted in the 1950s was the *Grampus*, of similar appearance to the *Lamprey*, with a body of steel plate and three piece side drop doors. A considerable number of these vehicles were introduced, built by a number of manufacturers, and both vacuum fitted and unbraked examples were constructed. One special feature when built was the fixing of cast nameplates, but the majority are now removed. ZBO No DB986348 is illustrated built in 1956 by the Cambrian Wagon Works. *Colin J. Marsden*

ZBA DC988600 *Carp.* A number of older short wheelbased ballast wagons are now up for replacement and during 1983 BREL Shildon produced No DC988600 as a modern air braked replacement for such as the *Grampus*. The vehicle painted in new departmental colours and was photographed at Eastleigh in May 1983. It was constructed under lot No 4032 and has a tare weight of 11,000Kg. Note the inclined ends, thus preventing ballast getting on to the draw gear. *Colin J. Marsden*

ZCV DS61972 *Tunny.* SR designed, but Birmingham Railway Carriage & Wagon Co produced *Tunny* wagons were introduced from 1928 and have a wooden body with two drop doors on each side with a carrying capacity of 20 tons. These were some of the first vacuum fitted ballast wagons and this design was later advanced into BR days with steel bodied vehicles of similar type. No DS61972 is seen at Hither Green. *Colin J. Marsden*

ZCV DB987104 *Plaice.* In 1977 BREL Horwich converted four *Plate* wagons into a new departmental ballast type entitled *Plaice* and a production run of 200 vehicles of the design followed during 1979-80. The wagons converted from *Plate* underframes have new uppers with drop sides and an opening at the top which, when lowered provides steps for staff. A maximum load of 22.5 tons can be carried and various detail differences can be found on some vehicles. No DB987104 illustrated has 'hybox' axleboxes and is painted olive green. The number range for these conversions is DB987100-303. *Colin J. Marsden*

Right, top to bottom:
ZDV KDB732953. Under the TOPS wagon code system ZD is for general material carriers of virtually any type and able to carry any weight. ZDV No KDB732953 is operated by the CS&TE engineers and was formerly a *Tube* wagon with wooden sides and steel ends. This type of vehicle is often used to carry cable troughing to engineering sites. The vehicle illustrated was built in 1955 at BR Darlington works. *Colin J. Marsden*

ZDV DB741897. This 12 ton former *Pipe* wagon, now operated by the CCE department, is classified as a general materials carrier and was used for barrier purposes on nuclear flask trains during 1983. It was photographed in sidings outside the Windscale Plant. No alterations have been made to this wagon since its revenue earning days. No DB741897 was built in 1961 at Wolverton. *Colin J. Marsden*

ZDV DB726378 *Dace*. A conversion was made during the early 1980s to nearly 200 OUV *Shockhood B* vehicles used for the conveyance of coil steel, these were rebuilt at New Cross Gate into *Dace* ballast wagons having the high-bar removed and additional steps on diagonally opposite corners fitted. Painted in olive green No DB726378 stands at Woking. All vehicles of this type have roller bearing axles.
Colin J. Marsden

ZDV ADB769678. The ZDV code is also given to former vans now in departmental traffic as general material carriers. No ADB769678 is used to transport sand in the Toton and Derby area and is a rebuild of a former *Vanfit* built in 1956 at the BR works Darlington. No alterations have been made to this wagon except for the additional lettering in white on the side. *Colin J. Marsden*

Above: **ZDV CDE260871.** A number of suitable flat wagons are in departmental stock for the transporting of stores between various BREL workshops and BR running depots. Vehicles so used are under the control of BREL and have the prefix 'C' to their number. This former *Lowmac EP* is used to transport locomotive bogies to and from BREL Doncaster works and is provided with a special central track section. *Colin J. Marsden*

Below: **ZDV ADS39602.** Operating with the CM&EE power supply section are a number of flat wagons, able to transport electrical transformers etc to engineering sites. This vehicle formerly SR *Conflat C* is now allocated to Horsham for this use. Few alterations have been made to this wagon since its revenue earning days. *Colin J. Marsden*

Above: **ZDV ADB453147.** This former *Lowfit* of all steel construction built by BR Shildon works was introduced in 1959 and is now operated by the CM&EE RSE section and used to convey EMU/coach bogies between depots. Again no major alterations were needed for its present use. *Colin J. Marsden*

Below: **ZDO DB453126.** Although there is only a passenger network on the Isle of Wight a small number of departmental vehicles are allocated to the system for essential permanent way work or emergencies. Unfitted former *Lowfit* No DB453126 was in use on the island during the electrification work in the 1960s but has seen little further use. It retains a central bolster and is presently parked out of use at Sandown. *Colin J. Marsden*

Above: **ZEO DP101453.** The Isle of Wight railway plays host to an amazing array of rare and in some cases otherwise extinct types. This unfitted ballast hopper with centre chute is formerly a privately owned *Ironstone* hopper transferred to the island in the mid-1960s. It is painted in bauxite brown with white lettering. Note the unique step and grab rails on the side by the hopper door wheel.
Colin J. Marsden

Left: ***ZEV DB992580 Catfish.*** A number of different designs of small four-wheel hopper wagons are in traffic with the CCE department for the conveyance of stone to engineering sites. The *Catfish* type, a steel bodied design, was introduced between 1955-58 mainly for LMR operation, but today can be found all over the network. This vehicle carries 19 tons of stone which can be dropped only via a central chute. All were built by Metropolitan Cammell. Both oil and roller bearing axle boxes can be found on this type. *Colin J. Marsden*

Below left: ***ZFV DB993623 Dogfish.*** A slightly enlarged version of the *Catfish* was introduced from 1956 titled *Dogfish* these have a maximum load of 24 tons and discharge doors in both central and side positions. Various batches of *Dogfish* have been produced all fitted with vacuum clasp brakes. No DB993623 illustrated was built in 1960 by BR at Shildon to lot No 3255. *Colin J. Marsden*

Above: **ZGV DE276604.** Former LNER *High* No E276604 is now in departmental operation as a general materials carrier with the CCE department and when photographed in 1983 was at Westhouses depot. This wagon is in almost original condition with small side drop door and able to carry up to 13 tons. The livery is dark green with white lettering. *Colin J. Marsden*

Right: **ZGV DE281570.** A number of former steel bodied *High* wagons are now under the control of the departmental board for use by the engineers department. This LNER built example is now operated on the SR as a general materials carrier for CCE operation. When photographed at Surbiton in 1982 the livery was olive green with white lettering. *Colin J. Marsden*

Below right: **ZGV ADB740224.** Another ex-revenue vehicle operating for the CM&EE power supply section is this former *Pipe* wagon No B740224 constructed by BR works at Derby in 1949 to the original LNER design of vehicle. It is now used to carry plant to and from electrical sub-stations and the SR power supply base at Horsham, where the vehicle was photographed in 1983. *Colin J. Marsden*

Above: **ZGV ADS1756.** This very old vehicle dating back to the 1920s is an SR eight-plank open fitted with vacuum brakes. The central drop door has two side hinged gates above, and the wagon has a total carrying capacity of 12 tons. It is now operated by the CM&EE RSE section and was photographed at Fratton in 1983 with light olive green livery and white lettering. *Colin J. Marsden*

Below: **ZGW ADB715030.** This former *Ferry Open* conforming to the RIV operating standards is now in departmental use again with the CM&EE RSE section. It was built by BR at Lancing in 1957 as one of 20 wagons to lot No 2851 and has a carrying capacity of 21 tons. This is a medium wheelbase vehicle of 14ft 10in and is therefore non-standard in many respects. It is fitted with vacuum brakes and a single through air-pipe. The RIV tethering points can be seen on the solebar. *Colin J. Marsden*

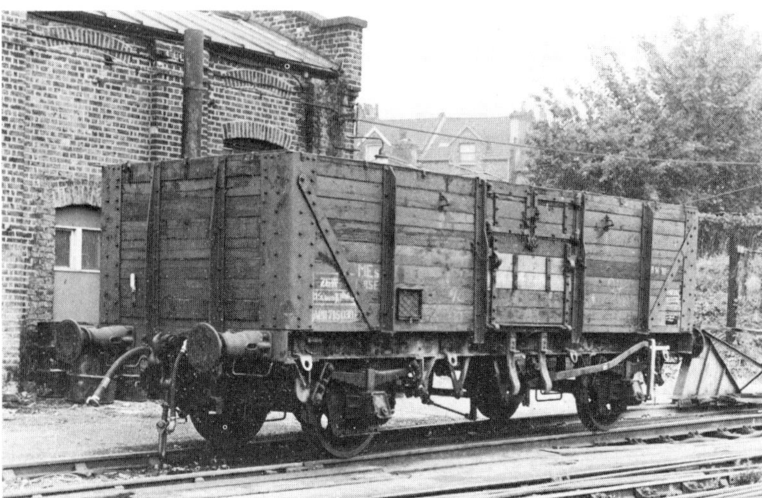

Above: **ZIB ARDC96705.** At a number of maintenance depots throughout the BR network high capacity cranes are positioned for use in case of emergency, derailment or occasional routine maintenance operations. The unit illustrated is based at Motherwell and numbered ARDC96705 and was manufactured in 1960 by Cowans Sheldon; it is a diesel-hydraulic crane capable of lifting 76 tons. In operational order it weighs over 130 tons and works with runner No ADB998537. Livery applied is yellow. *John Tuffs*

Below: **ZIP ARDC95204.** A number of steam powered breakdown cranes are still in operation. The one illustrated here was manufactured by Cowans Sheldon of Carlisle in 1930 as a 36ton capacity unit, being rebuilt for 50ton lifting in 1938 and operates from Haymarket depot. This crane weighs over 95tons when in working condition and combines with match/runner No ADM299855. When photographed in 1982 this set was painted in red livery. *John Tuffs*

Above: **ZIV ARDC96201.** There are only two breakdown cranes in operation on BR powered by a diesel-electric engine and these are both Cowans Sheldon 76ton capacity sets allocated to the SR at Ashford and Wimbledon. The Wimbledon based crane is illustrated here at Woking recovering some damaged ballast wagons following a derailment. Note the balance jacking arms protruding from the base of the crane and the radio operator standing on the adjacent track giving lifting instructions to the crane driver. This crane is unusual as its number is not carried on its bulk but on one of the escort wagons. *Colin J. Marsden*

Right: **ZIP ARDC95221.** Most regions have a relief breakdown train or crane which can be called upon if an emergency arises and the usual equipment is out of commission. The LMR relief crane is Cowans Sheldon 45ton capacity steam crane No ARDC95221 of 1943 vintage, which works coupled to jib match/ runner No ADM396008. During 1984 the crane and runner are stored awaiting attention at BREL Doncaster and painted in red livery. *Colin J. Marsden*

Below right: **ZIP ADRR95212.** Some Ransomes & Rapier steam cranes are still in operation, with lifting capacities of 34-45tons. Most of these cranes are scheduled for early retirement. No ADRR95212 illustrated is allocated to Old Oak Common depot and was built in 1940 with a 45ton lift capacity. It is painted in black livery and operates with match wagon No ADW150442. Note the chimney tipped downwards behind the rear of the crane boiler. This crane was taken out of use during 1983 when a new 76ton diesel-hydraulic unit was delivered. *Colin J. Marsden*

Below: **ZIB ARDC96713.** The latest batch of breakdown cranes were introduced between 1977-80 when Cowans Sheldon produced six diesel-hydraulic sets with telescopic jibs requiring no runner vehicles. These cranes are allocated to BR, ED, OC, SP, HM and KD depots and weigh 117 ton in full working order, are 68ft in length, and can be hauled at up to 75mph. No ARDC96713 illustrated belongs to Wigan. *Keith Grafton*

Above: **ZJV DB989637 *Mermaid.*** This type of ballast hopper with side tipping facility designated *Mermaid* is of GWR design, but all examples now in service were built in BR days. They are unique in the departmental ballast fleet as the complete body tips to discharge its 14ton cargo. No DB989637 shown here at Willesden was built in 1961 by Metropolitan Cammell. *Colin J. Marsden*

Below: **ZKO DB386737.** This vehicle was originally built in 1958 by BR at Derby works as an ironstone tippler to lot No 3091. It remained in this use until the early 1980s when it was transferred to the departmental fleet as a general materials carrier and mainly used to convey sand, cement and ballast to engineering sites. No external alterations have been carried out to this vehicle. *John Tuffs*

Above and right: **ZLV DB992419, DB992453 *Herring.*** The BR 20.5 ton ballast hopper with a centre chute is named *Herring* and numbered in the DB992381-DB992480 range. All were built by Metropolitan Cammell in 1953/4 and fitted with vacuum brakes. When constructed the fleet was destined for LMR operation, however over the years they have seen operation throughout the BR network. The upper plate shows No DB992419 allocated to the Isle of Wight, whilst the illustration to the right is of No DB992453 operating on the LMR under the control of the CCE at Crewe. One end of these vehicles has a verandah where the chute control wheel is situated and the other end houses the vacuum br cylinder. *Both: Colin J. Marsden*

Below right: **ZMV DB992284 *Mackerel.*** Another design of small four-wheeled hopper with centre chutes is the *Mackerel* with a maximum carrying capacity of 17.5 tons. The present fleet is numbered in the DB992247-DB992380 range. All vehicles were constructed by Metropolitan Cammell during 1951/2, again this type was intended for LMR operation but can be seen throughout the BR network. *Colin J. Marsden*

Above: **ZOO ADM15/6½.** The TOPS classification ZO is for travelling cranes, mounted on a fixed wheel chassis, ranging from 3 to 30 tons lift capacity and constructed by a variety of manufacturers. ZOO No ADM15/6½ illustrated, is one of the very few small Cowans Sheldon units left in use, with a lift capacity of just 6½ tons and is, of course, completely hand operated. It is allocated to the CCE at Lancaster and when photographed in 1982 was at Horton-in-Ribblesdale. *Colin J. Marsden*

Below: **ZOR ARDC96101.** For use by the CM&EE power supply section on the SR based at Horsham is Cowans Sheldon 30tons diesel-hydraulic crane No ARDC96101. This crane is painted in yellow livery and operates with match wagon DB998531. The unit is unusual as the yellow livery does not extend to the underframe of the crane, which is painted black. This crane is normally used in connection with electrification renewals. *Colin J. Marsden*

Above: **ZOP ADRS95000.** There are still a handful of old steam cranes in operation with the Civil Engineering departments. This example, well over 50 years old, is allocated to the LMR and operates in the Derby area, it has a lift capacity of only 5-6 tons and weighs 21 tons when in working condition. Its match wagon ZSV No ADB33342 is converted from a former *Plate* wagon and has small tool lockers at one end and a jib support in the middle. *Brian Morrison*

Centre right: **ZOR DRT81124.** The Civil Engineering Department on the ER, based in East Anglia, uses this 16ton lift capacity Taylor Hubbard crane for some of its operations. The livery is yellow with wasp warning ends. This crane operates with two match/escort vehicles, one as a supplies wagon and the other as a jib runner, both are classified as ZSR. *Barry J. Nicolle*

Below right: **ZPV DB773398.** The ZP code is allocated to staff or staff/dormitory vehicles usually operating with mechanical plant. No DB773398 was formerly a *Vanfit* built in 1957 at the BR works at Wolverton. It now sports two end and one side window, cooking and heating facilities and side steps and provides accommodation for staff operating the large Armfield crane. The livery applied is olive green with white lettering. *Colin J. Marsden*

Left: **ZPV DW17931.** Many pre-Nationalisation vehicles are still in traffic providing accommodation for staff. This former GWR brake van is now used as a staff sleeping and messing van for WR CCE crane No 81549. When photographed in 1982 the livery was green with yellow warning ends. *Colin J. Marsden*

Centre left: **ZQV DW150344.** Another former GWR vehicle still in departmental service is *Fruit C* No W2856, now used as a tool van for the CCE department based at Woking. This vehicle has seen little alteration since its revenue earning days, and still retains its two double doors on each side. When photographed at Stewarts Lane depot in 1981 the livery was green with white lettering. *Colin J. Marsden*

Below left: **ZRV B749100.** The TOPS code ZR is allocated to stores vehicles and in many cases tank wagons. No B749100 is now used as a bitumen tanker for the CCE department. It was formerly a Beer tank built by BR at Derby in 1950 for use by McEwan Ltd and was one of only two built under lot No 2074. The vehicle is fitted with vacuum clasp brakes and painted grey (under a heavy coat of bitumen). *Colin J. Marsden*

Right: **ZRV ADW3052.** This six-wheeled vehicle was built by the GWR as a milk tank for operation in the West Country and when made redundant in the late 1960s was transferred to the departmental fleet as a water carrier for the Old Oak Common based steam crane. It has now been taken out of service.
Colin J. Marsden

Centre right: **ZRV ADB998990.** There are few purpose built departmental tank vehicles in operation, but during 1962 a fleet of 38 four-wheeled creosote tankers were constructed by Charles Roberts of Wakefield under lot No 3401. During the 1970s some of these vehicles were transferred to the CM&EE department to provide additional water carrying facilities for steam breakdown cranes.
No ADB998990, now painted in red livery with white lettering, is allocated to Stewarts Lane. *Colin J. Marsden*

Below: **ZRP DB749604.** In 1951 a fleet of 10 four-wheel tank wagons was constructed for BR by Cravens Ltd to carry Ethyline Oxide, a chemical used in the manufacture of plastic film, and numbered B749600-9. Several remain in service today as water tank wagons for the DCE department in Sheffield. Four are shown here at Leeds painted in yellow livery.
Gary Grafton

Above: **ZRV ADM209670.** A vast number of pre-nationalisation short wheelbase vans are in traffic under the departmental flag for the conveyance of general stores between depots, stations and works. Former LMS 12ton sliding door van No ADM209670 is now operated by the CM&EE to transport loco sand. This 1920s built vehicle has a wooden body with metal ends. On the side it bears the legend Llandudno Junction, but when photographed in 1983 was at Westhouses. *Colin J. Marsden*

Left: **ZRV CDM515438.** Another former LMS vehicle is No CDM515438 which is operated by BREL, according to its prefix number, and is used to transport battery accumulator cells and thus titled *Cell Truck*. This vehicle is again of wooden construction mounted on a steel frame and has steel ends. Large wooden sliding doors are provided on each side. This vehicle is allocated to BREL Wolverton and operates between here and a number of the larger carriage depots around the network.
Colin J. Marsden

Below left: **ZRV ADB766234.** A more modern BR built van that is now in departmental traffic is this *Vanfit* built in 1956 by BR at Darlington. It is now used by the CM&EE as a stores vehicle and based on the SR. The livery is olive green with white lettering and no alterations have been made for its use as a departmental carrier.
Colin J. Marsden

Right, top to bottom:

ZRV DW100682. This rather unusual stores vehicle was formerly a 1939 built GWR sleeper wagon and has been in use for many years as a concrete stores carrier allocated to the WR concrete depot at Taunton, from where it normally conveys concrete cable troughing or sleepers to engineering sites. This type of vehicle design dates back to 1894. The illustrated example has acquired wooden plank sides in recent years and is fitted with clasp vacuum brakes. Photographed at Woking in 1982, livery is green with white lettering.
Colin J. Marsden

ZRW ADB998050. Four purpose built departmental vehicles were constructed by BR Ashford in 1957/8 for the conveyance of English Electric 6K engines, as fitted in the SR diesel electric multiple units. The wagons are 13ton flats which have a small locker facility at each end, the engine sits on a fabricated bed and a steel lid is placed over the top to prevent water or other damage. These vehicles operate between Eastleigh and St Leonards and are numbered ADB998050-53.
Colin J. Marsden

ZRV Denparts & Conflat. Two types of departmental ZR vehicle are shown in this illustration. Nearest the camera are two *Denparts* vehicles carrying small items from one BREL works to another, such as electrical components. Coupled behind the *Denparts* wagons are three former *Conflats* now used to transport bogies between BREL workshops.
Colin J. Marsden

ZRO ADM707111. Another *Conflat* that has seen further use is this former LMS vehicle No ADM707111 which for many years has carried a demountable liquid tank in the Derby, Burton and Coalville area. The tank was formerly a road vehicle used to convey locomotive anti-freeze to depots. When photographed in 1983 at Coalville it was parked out of use. The last maintenance date recorded on the vehicle was 11 January 1965.
Colin J. Marsden

Above: **ZSV DM475530.** The TOPS code ZS is given to a wide number of vehicles, all are four-wheeled barrier or runner wagons. This former LMS three plank open No M475530 is now used as the jib runner and equipment carrier for DCE crane No RM297D-15/6½ allocated to Lancaster. This wagon has a wooden body with vacuum push brakes and was constructed in the late 1930s. *Colin J. Marsden*

Below: **ZSO DB731565.** Now used as the jib runner for crane No DRF81315, this wagon was originally a 22ton *Tube* constructed in 1953 at BR Wolverton Works. Most of the body sides have been removed except for shallow fixed sides and ends with a steel jib support in the middle. The wagon is not continuous brake fitted, and is painted yellow and black. *Colin J. Marsden*

Above: **ZSR DB998513.** A number of purpose built crane match/runner wagons are in service classified ZS. Often two vehicles are allocated for use with one crane, one is the jib runner, the other for tools and equipment. No DB998513 illustrated has a frame to carry oil drums and packing equipment. This wagon is not in its correct active position in this photograph, but is awaiting maintenance at Stewarts Lane depot. *Colin J. Marsden*

Below: **ZSR DB904062.** The Wimbledon East breakdown train equipment carrier, seen here covered with a tarpaulin, is No DB904062, which is a rebuild from a 21 ton *Lowmac MR* of 1951 vintage, built by P. W. McLellan as one of 40 vehicles under lot No 2276. The wagon has been rebuilt with sides and equipment storage cases. This vehicle usually carries spare lifting tackle and trolleys. Livery is red with white lettering. *Colin J. Marsden*

Left: **ZSV ADB530597.** After orders were placed for 1972 design high-density EMU stock with low-level *Tightlock* couplings, a method of transporting them around the system was sought, as their couplings were incompatible to standard stock. To overcome this problem some former *Conflat* wagons were converted at BREL York to act as match vehicles, with low-level *Tightlock* couplings at one end, and conventional drawgear at the other. Two match wagons are illustrated here coupled together, with the conventional drawgear outward. *Colin J. Marsden*

Below left: **ZSW ADB452732.** Coupling problems also prevail when the Waterloo & City line cars require movement on the main lines to and from depots, as these vehicles are fitted with *Ward* couplings, again not compatible to conventional drawgear. Several match wagons exist with Waterloo & City railway couplings at one end. The match end of the vehicles have their buffers removed and a *Ward* coupling fitted. ZSW No ADB452732 illustrated, is converted from a 13ton low goods. Note the vacuum pipes laying across the vehicle, these are laid through the Waterloo & City car enabling the vacuum brakes to be operative each side when in transit. *Colin J. Marsden*

Below: **ZSO ADS59045.** Coupling problems were also encountered on the Isle of Wight system, when the converted tube stock was transferred to the network in 1966. To solve this, two match wagons were converted at Ryde works with low level *Ward* couplings at one end and conventional shackle couplings at the other. These two vehicles are now out of use at Sandown and are wooden bodied and framed lows built just after the turn of the century. Note the LB&SCR axle box nearest the camera. *Colin J. Marsden*

Left: **ZSV TDB701983, ZSX DB451897.** For a number of years some of the smaller shunting loco types have caused problems with track circuit equipment, where in places their presence has not been recorded on the signalling panel, this has often been attributable to their light weight and short wheelbase. To overcome this in a number of areas match wagons are provided, these are semi-permanently coupled to locomotives and carry a repeat of all end connections. The upper plate shows No TDB701983 which is fitted for vacuum brake operation and is a converted *Conflat A*. The illustration on the left shows No DB451897 a dual braked vehicle which was formerly a 13ton low goods.
John Tuffs/Colin J. Marsden

Below: **ZSV ADS70188.** Passenger coaching stock fitted with steam heating equipment often require 'pre-heating' before entering service. If a locomotive is not available a boiler 'pre-heating' van is sometimes provided which is fitted with a standard steam heating generator. No ADS70188 is a purpose built vehicle introduced during the 1960s, built on the underframe of former mineral wagon No S40398. This vehicle looking similar to the now extinct horsebox vans stands at Old Oak Common. It is allocated to the WR and is usually used in the Oxford area. *Colin J. Marsden*

Left: **ZTR DS55560, ZTO DS55710, ZTV ADS455.** A considerable number of ex-traffic brake vans have passed into departmental use. These three illustrations all of ex-SR vans, are each fitted with a different brake type, all the vans conform to the same basic design although detail differences do exist. The top plate of No DS55560 is not a continuously brake fitted vehicle but has both vacuum and air through pipes. The middle plate is of an unfitted/unpiped van No DS55710 which operates on the Isle of Wight. The third illustration of by far the oldest van No ADS455 is fitted with conventional vacuum brake gear. All three vans are painted in olive green livery. No ADS455 which is stationed at Guildford is the only surviving 15½ton Brake Van in service.
All Colin J. Marsden

Bottom left: **ZTO KDB951758.** Today a number of BR designed brake vans are in departmental traffic operating with various departments and often for specific purposes. BR standard unfitted van No KDB951758 built by BR at Darlington in 1952 is now used by the S&T department based at Doncaster, therefore having a K prefix to its number. *Patrick Kingston*

Above: **ZTP (ZTQ) DM730305.** This former LMS type brake van is now used as an escort/brake van for one of the LMR gauging trains. It does not have a continuous brake but is fitted with a through air pipe (this wagon should, therefore, be coded ZTC but was incorrectly stencilled on its last overhaul. When seen in 1982 the livery was olive green with yellow ends. At the other end of the gauging train is a former GWR *Toad* van. *Colin J. Marsden*

Below: **ZTO DM732157.** This former LMS van now looking rather worse for wear, is non-brake fitted and some body panels have been removed for observation purposes, it now operates with the BRB bitumen spray train. All windows in this vehicle have mesh guards and the majority of internal fittings are removed. The livery is bauxite brown and white, with white lettering on the sides. *Colin J. Marsden*

Bottom: **ZTO DM732332.** Another converted former LMS brake van is this application as match wagon for *Tightlock* fitted EMU stock. One end of the vehicle retains conventional drawgear while the other is fitted with a low level buck-eye (*Tightlock*) assembly. The two LMS vans coupled either end of this six car formation of Class 508 stock are allocated to Hall Road, Liverpool, and have been used to convey Class 508 units from the SR to their new home in Liverpool. *Colin J. Marsden*

Above: **ZUV DS62860 *Shark*.** Plough fitted brake vans were introduced to operate with ballast trains where the ploughs could be lowered to spread the ballast previously dropped from hopper wagons. This view is of an SR designed van constructed in the 1940s by the SR at Ashford. The wagon is fitted with vacuum and hand brake equipment and is still in service. Photographed at Doncaster in 1982, livery is olive green with white lettering. *Colin J. Marsden*

Below: **ZUV DB993923 *Shark*.** Apart from the SR ploughs a number were also constructed by the LMS and LNER all of similar design. BR carried on the LMS design and in 1962 introduced the ultimate in the *Shark* series when 20 vehicles fitted with oleo pneumatic buffers and roller bearings, built by the Central Wagon Company of Wigan and numbered DB993921-40 entered service. No DB993923 is illustrated bearing its home location as Nuneaton Abbey Street. *Colin J. Marsden*

Above: **ZVV CDB737866.** This TOPS code ZV is allocated to a variety of miscellaneous vehicles, all used for carrying purposes. This former *Conflat A* now classified as ZVV and allocated to BREL Derby Litchurch Lane works, is used for bogie transportation. No modifications were carried out to this vehicle for its present use. No B737866 was built by BR Ashford in 1959 as one of 400 vehicles under lot No 3107. *Colin J. Marsden*

Below: **ZVV ADB460168.** This former 13ton *medium* goods built in 1952 at BR Ashford is now allocated for the use of H. Pooley and Son Ltd as a weight test vehicle and based at Norwood Junction. The wagon carries specific weight blocks and travels around the network checking the operation of the various weigh tables. Apart from its load, and side lettering, this vehicle is in original condition. *Colin J. Marsden*

Above: **ZVV ADM476401.** This former LMS three plank open service wagon now has a special semi-permanently fixed cargo — a winch. It is operated by the LMR CM&EE department based at Crewe and used for the pick up of lineside equipment, often wheelsets. The wagon body is black and the winch yellow. Photographed at Willesden Yard, in 1982. *Colin J. Marsden*

Below: **ZVV ADM904668 *Lowmac EU.*** A considerable number of former revenue fleet low wagons are now under the control of the departmental board, used for a variety of duties. This *Lowmac EU* now operated by the CM&EE is an equipment carrier and was used in 1982/3 to carry electrical sub-station assemblies between Horsham and engineering sites. The wagon was built in 1955 by the Derbyshire Carriage & Wagon Company. Photographed at Horsham. *Colin J. Marsden*

Left, top to bottom: **ZVP ADB904152 *Lowmac AB.*** Yet another vehicle operated by the CM&EE for the transport of bogies is this *Lowmac AB* which was one of 10 vehicles constructed by BR Shildon in 1960 under lot No 3254. This vehicle has a maximum load of 21.5 tons and is fitted with a vacuum through pipe and also roller bearings. The livery in 1982 was olive green with white lettering. *Colin J. Marsden*

ZVV ADB904144 *Lowmac MU.* The number of CM&EE operated flats seems to be endless, this *Lowmac MU*, incorrectly labelled *Lowmac SE*, is yet another bogie or wheel transporter wagon. It was built at the BR workshops Swindon in 1959 as one of 10 wagons under lot No 3101. It is fitted with vacuum braking and has roller bearing wheels. When photographed in 1982 at Guildford it was painted in olive green livery with white lettering. *Colin J. Marsden*

ZVP DB998017 *Flatrol ED.* Some of the Flatrol or Lowmac series vehicles are operated by the CCE department for the transportation of ballast machines to engineering sites. This *Flatrol ED* is worked by the LMR CCE and photographed at Carnforth in early 1983. By the state of the well part of the vehicle it is suspected that it has been recently used to transport a mechanical digger. No DB998017 was built by BR Lancing in 1959. *Colin J. Marsden*

ZVW ADB900118, 909 6012-1 *Flatrol EAC.* There are few departmental vehicles that conform to UIC or continental standards, but this *Flatrol EAC* is still in traffic, operating with the CM&EE department. When photographed at Salisbury it was carrying Talyllyn Railway 0-4-0 WT *Dolgoch* as part of a nationwide exhibition train. This vehicle has vacuum brakes and through air pipes. *Andrew French*

Above: **ZWA DB968500.** Due to the adverse weather conditions that prevail in Scotland, in 1980 it was decided to introduce a snow-blower based at Inverness. The vehicle was constructed under lot No 4000 and numbered DB968500, being built by Atkinsons Beilhack in West Germany. The blower is powered by a 450hp Deutz engine and can clear 6,000tons of snow per hour. For traction the blower is powered by a Deutz 200hp engine with hydrostatic transmission. The unit is seen inside the shed at Inverness. *Derek Porter*

Below: **ZXO DM153163.** The correct gauging of the BR network is of prime importance and a number of gauging trains operate throughout the system. The LMR unit, illustrated here is built on a former brake van body and consists of a number of feeler gauges positioned at various points around the width and height of the vehicle. The gauging van has no brakes (not even a handbrake) and is therefore semi-permanently coupled to its two escort vehicles. *Colin J. Marsden*

Above: **ZXO DW35883.** Operating with the gauging feeler van is this former GWR *Toad* brake van, which now provides brake facilities for the train and accommodation for staff. It is painted in departmental olive green with yellow ends. It states on the sides that the van must not be used as a conventional brake vehicle and is not to be loose shunted. The LMR gauging train is photographed at Willesden during 1982. *Colin J. Marsden*

Below: **ZXV ADB936413 *Leapfrog.*** A *Plate* wagon was rebuilt in 1964 to accommodate a 10ton demountable petrol tank. The *Plate* chassis was completely rebuilt with lower decking, the top of the solebar being left proud of the decking on to which were mounted three pairs of retention clips, these located into the base of the tank to hold it into position. A suitable road lorry was also constructed. The idea behind the scheme was to build a fleet of rail wagons, to load them at an oil refinery, operate them as a block train to a rail head where lorries would take the containers direct to the customer. The idea however was not adopted, and the wagon was eventually taken into departmental service at Derby Technical Centre where it can still be found, now painted in black livery. *Colin J. Marsden*

Above: **ZXX RDB787319.** Apart from the considerable number of former passenger vehicles operated by the Railway Technical Centre at Derby a few four wheel vehicles can also be found. This former BR ferry van No B787319 built in 1963 at BR Ashford, is now part of the vehicle/track interaction test train, it is fitted with dual brakes and is labelled as an auxiliary testing vehicle. The livery is departmental red/blue. Some alterations have been made to the side and door panels. *Colin J. Marsden*

Below: **ZXR RDC460000.** This interesting purpose built vehicle for the Research & Development section at Derby is an optical structure gauging vehicle, used for the measurement of clearance profiles. The wagon painted in matt black livery is unfitted but piped for both air and vacuum brake equipment and fitted for an electric train supply. When in operation it should be formed between Laboratory cars Nos 17 and 18. It is seen in the yard at the Technical Centre whilst being fitted out in 1983.
Colin J. Marsden

Above: **ZXO RDB999900.** During the 1970s when Derby Technical Centre formed the Tribometer test train to investigate adhesion problems, a COB-AB van of a recent build was transferred into departmental stock and rebuilt as an adhesion measuring vehicle. It is not conventional brake fitted but has a through brake pipe (air), however test brake equipment is fitted and often lightly applied when the vehicle is in motion for testing purposes. The van is painted in Research blue/red livery and is usually kept at Derby. Modified wheel sets have also been fitted. *Colin J. Marsden*

Below: **ZXA RDB511023.** This rather interesting vehicle operated by the Research & Development Section at Derby is titled HSFV-1, and operates in connection with the vehicle/track inter-action group. It is a rebuild of a container flat and now carries rail packs for extra weight. The vehicle is mounted on an experimental underframe and fitted with air brakes and a through vacuum pipe. Livery is rail blue underframe with yellow ends. *Colin J. Marsden*

Above: **ZZV ADE330977.** A sight that many enthusiasts will have noticed are the snowploughs stabled around the system. Under the TOPS classification these are 'ZZ' vehicles. These ploughs are not powered but are hauled/propelled by diesel locomotives and it is usual for the plough to be coupled at either end of one or two locomotives when required for use. A number of ploughs are rebuilt from ex-steam locomotive tenders, while others are purpose built. No ADE330977 illustrated is a former tender fitted with a medium sized plough. It weighs 30 tons, has a maximum speed of 45mph and is 28ft in length. *Michael J. Collins*

Below: **ZZV ADS70229.** Eight former 'Schools' class tenders were converted into independent snowploughs in the late 1960s and fitted with vacuum brakes. These are a large plough and have a retractable plough head, at the rear of the unit there is a small verandah where a handbrake and control wheels are positioned. Note the powered marker lights fitted on the lamp irons on the front. These eight ploughs numbered in the ADS70210/1, 70224-9 range are all allocated to the SR, weigh 32 tons, can be hauled at up to 45mph and are 28ft in length. No ADS70229 is photographed at Redhill.
Colin J. Marsden

Left: **ZZV ADS70227.** Rear end of SR independent plough, rebuilt from former 'Schools' class tender, showing the handbrake control wheel, plough control handle and entry position into the plough body. Underframe mounted equipment consists of a screw coupling, and vacuum pipe. Buffing is provided by blocks, not buffers. No ADS70227 is allocated to Salisbury but was photographed at Reading in 1982. *Colin J. Marsden*

Left and below: **ZZV ADB965238, ADB965239.** The most numerous fleet of independent snowploughs is the 44 strong BR standard type constructed from 1965, with three additional units built in 1970. These ploughs weigh 32 tons and are fitted for vacuum brake operation, can operate at 45mph and are 28ft in length. The number range is ADB965203-243, ADB965307-9 and they are allocated to depots throughout the country. Construction was carried out by the BR works at Cowlairs, Eastleigh, Swindon and Derby. The two LA allocated ploughs are shown in the upper plate while the rear view below is of the plough allocated to Tinsley. Control and handbrake equipment on these units are situated inside the plough body. *Colin J. Marsden*

Class 97 locomotives and ex locomotive stock

Former locomotives now in departmental stock. Machines classified 97xxx are powered vehicles, and can be used for traction purposes. Those carrying DB968xxx numbers are non-powered vehicles used for static purposes.

Traction Units

Departmental No	*Former No*	*Use*
97201	24.061	Derby Technical Centre loco
97250	25.310	ETHEL No 1
97251	25.305	ETHEL No 2
97252	25.314	ETHEL No 3
97402	46.023	Test loco
97650	PWM650	CCE shunter
97651	PWM651	CCE shunter
97652	PWM652	CCE shunter
97653	PWM653	CCE shunter
97654	PWM654	CCE shunter
97701	M61136	Battery engineers loco
97702	M61139	Battery engineers loco
97703	M61182	Battery engineers loco
97704	M61185	Battery engineers loco
97705	M61184	Battery engineers loco
97706	M61189	Battery engineers loco
97707	M61166	Battery engineers loco
97708	M61173	Battery engineers loco
97709	M61172	Battery engineers loco
97710	M61175	Battery engineers loco
97800	08.600	CMEE shunter Slade Green
97801	08.267	Railway Technical Centre remote control test loco
97804	06.003	S&T works shunter, Reading
97805	03.079	Isle of Wight CCE loco
Two shunters carry Internal User numbers for pilot duties		
PO1	08.173	Polmadie depot shunter
TS1	08.067	Tyseley depot shunter

Static Units (Ex locomotive stock)

D2991	—	Generator loco
ADB968000	D8243	ETH supply unit
ADB968021	84.009	CMEE load bank
83009	—	Power supply vehicle at Longsight

Below left and above: **97201 *Experiment.*** The Research Section at Derby often require to operate test trains, and to save the allocation of a traffic locomotive for this purpose, former Class 24 No 24.061, after its useful life was finished, was allocated to Derby to operate test trains when required. Another advantage of using a specific locomotive for test purposes is that any trial equipment fitted can be on a semi-permanent basis. No 97201 when not in use, can usually be found at either the Railway Technical Centre or Etches Park depot. The livery is departmental red/blue. The name *Experiment* has now been removed. *Derek Porter/Colin J. Marsden*

Below: **97250 *Ethel 1.*** A recent introduction into the departmental fleet has been three former Class 25s, which have been rebuilt to give an ETH supply to stock operating in Scotland usually hauled by steam heat fitted locomotives. The rebuilding has included the removal of some traction equipment and transferring the main generator output from traction to ETH use. The locomotives are allocated to Eastfield and painted in blue/grey coach livery with their ends painted in all blue. Ethel stands for Electric Train Heat Ex Locomotive and three names — *Ethel 1-3* are painted on the sides of Nos ADB97250-2. ADB97250 is illustrated, headed by a Class 37 at Cowlairs Junction. *Tom Noble*

Above and below: **PWM Class 97/0.** Between 1953-9 five Ruston & Hornsby 0-6-0 diesel-electric civil engineering shunters were introduced. These locomotives were built primarily for operation in engineering yards or occasionally on the main line at the location of major engineering work. Although the five locomotives are allocated to Radyr, Taunton and Reading, one of the fleet is often used at Swindon. They were originally numbered in the PWM (Permanent Way Machinery) series, but following the introduction of TOPS they became Class 97/0 Nos 97650-654. The upper plate shows No PWM651 now No 97651 at Radyr, while the lower plate illustrates No 97652 painted in yellow livery at Taunton. *Colin J. Marsden/Ian Gould*

Above: **97800 *Ivor.*** The SR CM&EE department have sizeable workshops at Slade Green where there is, on a part-time basis, need for a diesel locomotive to shunt stock. To save the cost of using a traffic locomotive the CM&EE acquired Class 08 No 08.600, an air braked example, and now operate it under the departmental flag as No 97800 *Ivor.* The livery is departmental blue/red and on the side by the name are four handpainted daffodils. The locomotive is otherwise in original Class 08 condition. *Colin J. Marsden*

Below: **97801 *Pluto.*** Following the withdrawal of Class 08 No 08.267 in March 1978, the Research Centre at Derby took over the locomotive for train control tests involving automatic train operation. A considerable amount of experimental equipment has been fitted to this locomotive including additional front lights and side indicators. No 97801 is allocated to Derby Technical Centre but usually operates on one of the test tracks. Livery is rail blue with white lettering. *Research & Development Section Derby*

Above: **97803.** Although the Isle of Wight railway is primarily an electric passenger network the system plays host to one departmental shunter, used to haul the Island's engineering trains No 97803 formerly Class 05 No 05.001 was used from 1966 until 1983 when repairs were needed. It was then replaced by No 97805 a former Class 03 No 03079. The Island shunter is allocated to Ryde but often kept at Sandown. *Colin J. Marsden*

Below: **97804.** For a number of years the Signal & Telegraph department at Reading has operated its own diesel shunter. Until 1982 this was provided by a Ruston 0-4-0 machine No 20, later 97020, but after this locomotive was found to be unsuitable for further use, a redundant Scottish Region Class 06 No 06.003 was drafted into departmental use and renumbered 97804. The locomotive retains rail blue livery and can usually be found in the S&T yard. The machine seldom operates to the nearby Reading diesel depot except for fuel or essential repairs. No 97804 is seen in the S&T yard during early 1984. *Sandra Fenn*

Above: **D2991.** The fleet of Ruston Class 07 locomotives built for use in Southampton Docks were all withdrawn from capital stock by 1977. No D2991, withdrawn in May 1973, still remains in use at BREL Eastleigh works as a mobile generator unit, however this locomotive has never, and is now unlikely to, be taken into departmental stock and allocated a departmental TOPS number. The external condition of the locomotive is now rather shabby but nevertheless it remains intact and serviceable and therefore worthy of an entry here. *Colin J. Marsden*

Below: **84.003.** Following the withdrawal in 1980 of the Class 84 25kV electric locomotives built by the North British Locomotive Company in 1960, two machines were transferred to Derby Technical Centre for further use. No 84.009 was rebuilt as a mobile load bank (unpowered) and painted in departmental blue/red livery. A number of former windows in the side were altered to louvres and several new ventilation openings were inserted assisting with the dissipation of heat. The second locomotive (illustrated here) No 84.003, is still at Derby Technical Centre for possible Departmental conversion at a later date and is the source of spare parts to the mobile load bank. *Colin J. Marsden*

Above: **D7076.** After the demise of the Class 35 'Hymek' locomotives on the WR two — Nos D7076/96, were taken over by the Research & Development Section, Derby, and used as dead load vehicles for track development tests, and based on the R&D test track at Old Dalby. By the early 1980s the locomotives were redundant and out of use, No D7076 was secured for preservation in 1983, but No D7096 remains at Old Dalby, now looking rather the worse for its long stand in the elements. *A. Colton*

Below: **97703.** A fleet of battery-electric locomotives were introduced between 1973/80, rebuilt from redundant Class 501 EMU stock, for use on electrification projects, involving semi-underground operation. Rebuilding work was undertaken on the first two vehicles at Birkenhead Depot and subsequently at BREL Doncaster. This included lowering the roofs, rebuilding the cab ends, sealing up all but two of the existing doors and windows, and the removal seats and replace them with battery cells. Normally two battery-electric locos operate together, livery is BR standard blue with yellow/black wasp ends. No 97703 is photographed at BREL Doncaster in 1980. *Colin J. Marsden*

Above: A number of redundant locomotives have seen further use as ETH supply vehicles for carriage sidings where fixed equipment is not available. In 1969 four redundant Class 15 BTH Type 1s were rebuilt for this purpose for use on the ER, numbered DB968000-3. Externally the locomotives were little altered except for the addition of ETH cables on the ends and the removal of some traction equipment. In 1984 only one unit remains No ADB968000, now allocated to Bristol, where it is used to pre-heat Mk III sleeping stock. No ADB968002 illustrated here is at Toton, awaiting scrap. *Colin J. Marsden*

Below: Four 'Toffee-Apple' Class 31/0 locomotives were also converted to ETH supply units for ER operation. These units were converted between 1976-80 and only remained in use for a short period being redundant by mid-1982 when all were withdrawn. Again little external modification work was necessary for their departmental use, except for the fitting of ETH cables on the nose ends and the removal of certain air/vacuum connections. No ADB968013 rebuilt from No 31.013 stands painted in green livery at Norwich depot during 1981. *Derek Porter*

Class 92 and 93x: Departmental unit stock

Class	*No range*	*Use*
920	920001	Test unit (former PEP)
930	001-013	De-icing units
930	70070-70087	De-icing cars
931	019, 020, 024	Stores units
932	975032	Test car MARS
933	055	Training school unit
935	056-057	Test units (former PEPs)

Other EMUs are in departmental service but are not allocated departmental classes.

Class	*No range*	*Use*
405	4623	Tractor unit
414	051-053	Tractor units
Unclassified		
—	975027	Bogie test car
—	975808/9	Chopper control test cars

Below: **Class 930.** Being that the SR is primarily an electric railway network, it is not surprising that several former EMU sets are now in departmental use. The 9xx classification system is awarded to these vehicles. The SR operates 13 two car de-icing units, the first two of which Nos 001-2 are rebuilds of former 2HAL motor cars and introduced in 1967. These carry tanks of de-icing fluid which is laid on the third rail by special shoes. During the autumn months the units are used for rail cleaning and fitted with sandite spray equipment. No 001 is seen at Wimbledon East Depot. *Colin J. Marsden*

Above and below: **Class 930.** Between 1976-80 many of the old de-icing units were replaced by former SUB motor coach conversions. These 'new' units were completely rebuilt with modified front ends and revised driving controls. Large spray tanks were fitted internally to carry the de-icing fluid and full staff facilities were also provided. The second generation de-icing units carry numbers 003-013 and are allocated to specific depots and in winter months operate timetabled services to ensure the 'live' rail is ice free. Units Nos 003-005 are also fitted for sandite spray duties in autumn months. Livery is rail blue with yellow warning ends. *Both: Colin J. Marsden*

Above: **Class 930.** A fleet of four trailer de-icing vans were introduced in 1959-60 converted from first batch augmentation SUB trailers, fitted with ep brake equipment and buck-eye couplings enabling them to operate between most post 1951 EMU types. The van carries a de-icing tank and the shoes on the bogies at either end lay the fluid. Three of the four vans remain in service. No ADS70050 illustrated was converted from car No S10392 from SUB No 4345. *Colin J. Marsden*

Below: The CM&EE depot at Strawberry Hill is the base for a tractor unit, formed of the two motor cars of 4SUB No 4623. The twin power is not officially in departmental stock but is used by the CM&EE for depot shunting purposes or operating special CM&EE duties around the region. During mid-1983 is was used to transfer test weight blocks from Shepperton to Alton. The unit is pictured at Strawberry Hill. *Colin J. Marsden*

Above: **Class 931.** Rather than use loco-hauled stores trains the SR operates a fleet of Stores EMUs, formed of redundant 2-car formations. The original sets formed of HAL type cars have now been withdrawn and the present fleet consists of unit 019 rebuilt from SR 2SAP; 020 formed of 1951 type EMU cars, and set 024 rebuilt from former SUB cars. Unit 019 is illustrated at East Wimbledon, painted in departmental green livery. *Colin J. Marsden*

Below: **Class 933.** The SR has operated a mobile classroom unit for a number of years, the present unit converted from 4SUB No 4387 entered service during the 1970s as number 055. It is owned and operated by the CM&EE department for the tuition of fitting staff. Each coach is converted into a classroom with a blackboard at one end and training equipment along the sides. The unit is allocated to Selhurst (SU) but can be found at any one of the SR EMU depots. Livery is BR blue. Note the nose end (SUB) jumper cables have been removed. *Colin J. Marsden*

Above and below: **Class 935, Class 920.** In 1971 the SR took delivery of the PEP prototype EMU sets which were operated under test and revenue earning conditions until 1976 when they were taken out of service. The two-car unit No 2001 was passed to the R&D section at Derby as early as 1974 and converted into a test bed for ac power equipment. The four-car sets Nos 4001/2 remained on the SR where a number of experimental fittings were added. Set No 4002 was taken to Derby Research Centre in 1979 and returned to the SR painted in executive livery where it again operated under test conditions relating to bogie and traction motors. Unit No 4001 was taken to Derby and stored at St Mary Yard. The upper view shows revised liveried No 4002 stored at Clapham Junction in 1983, while the picture below shows one car of unut No 2001 renumbered 920.001 alongside two cars of set No 4002 at Derby Technical Centre in 1979. *Both: Colin J. Marsden*

Bottom: Three former 2HAP units were taken into departmental stock at Strawberry Hill in 1982 to act as tractor units for moving various powered/unpowered vehicles around the region. The sets were also used for driving facilities on the chopper train control programme with test cars *Romeo* and *Juliet*. In early 1984 the units were engaged in the movement from BREL Eastleigh of the new Class 488 and 489 GLV stock for the Victoria-Gatwick service. Set No 053 formerly 2HAP No 6142 stands at Strawberry Hill depot. No external alterations have been made to these units although the majority of seats have been removed. *Colin J. Marsden*

Above and below: **Class 932.** Two redundant Class 501 cars now form part of the SR departmental fleet, Car No M61162 which is now ADB975027, and No M75165 now No ADB975032 *Mars*. Both vehicles have been extensively rebuilt as departmental stock No ADB975027 has been used for a number of bogie and traction motor experiments having some body alterations made to accommodate equipment and ventilation. No ADB975032 *Mars* is operated by the testing section at Strawberry Hill and fitted with some highly sophisticated trials equipment. It can be coupled to almost any EMU type train and provide full driving facilities. The car has been used in commissioning trials with Class 411/412 EMUs, and in 1983/4 with Class 488/489 Victoria-Gatwick push-pull sets. In the top plate No ADB975032 *Mars* is seen at Strawberry Hill, whilst the centre plate shows No ADB975027 stored at Wimbledon East depot. *Both: Colin J. Marsden*

Below right: The LMR introduced a two-car 'rail cleaning' unit in the early 1980s, to operate on the Merseyrail system, formed of redundant Class 502 MBSO vehicles, numbered ADB977017/018. The unit, normally based at Southport, remained in blue/grey livery although wasp ends were applied. It was withdrawn in 1983 and sold for scrap to King & Son of Snailwell. *Brian Morrison*

Track Maintenance Machines

Above: A number of the track maintenance machines that operate on the BR network are built and supplied by Plasser & Theurer of Austria, who have a sizeable British operation based at West Ealing, London, and where a number of machines are now constructed. This illustration is of a maker's plate attached to a type RCM-100 rail changing machine, built in 1979, and numbered 8948.
Colin J. Marsden

Below left: This type of machine is classified OBW by its builders Blasser & Theurer. It is a multi-purpose permanent way vehicle, with an Atlas jib unit on the rear and small storage skip capable of carrying 8 tons of cargo. These units are especially useful on small civil engineering sites where a full train is not required. No DX68200 is allocated to the SR and usually operates on the Central Division. *Colin J. Marsden*

Above and below: A number of special application machines operate on the Scottish Region, mainly in areas that are inaccessible by road or where there is little rail traffic. The plate below shows machine No 68028 a staff transporter and tractor unit, able to haul light permanent way loads to sites, and it can also convey 10-12 members of staff. The upper illustration shows machine No 68100, a multi-purpose permanent way motor vehicle also able to carry staff, haul a light load, normally one vehicle, and is fitted with a hydraulic grab. *Both: John Tuffs*

Above: This large machine built by Plasser & Theurer is a ballast regulator unit, which distributes ballast in the track centre, 6ft way (between lines) or in the cess (the outside edge of the track), it then brushes up all excess ballast and carries it in its own hopper. This type of machine is often found on the ER. *Colin J. Marsden*

Below: The Research & Development section at Derby have developed a stone blower or injection machine, based on the Plasser & Theurer 06-16-CTM machine No DW7258. This special set has been conceived to extend the period between track maintenance operations. The machine injects pre-determined quantities of new stone under each sleeper and after trafficing restores track to the correct level. The injection is made by blowing stone through a tube which probes the ballast adjacent to the sleeper. The machine is usually kept at Derby. *Colin J. Marsden*

Above and below: In 1984 26 of these USP5000C machines were in traffic allocated to all regions of BR, numbered in the range 77200, 77301-325. This is a universal ballast distributing and profiling unit, sweeping up unwanted ballast and storing it in an on-board hopper to redistribute as required. The machine is also fitted with a profiling arm which places the ballast in the correct position to the side of the track as well as underneath. The units weigh 42 tons and are powered by a 188kW engine. No DX77319 of ER is illustrated. *Both: Colin J. Marsden*

Above and below: Often when track relaying is undertaken some of the ballast can be used again but to ensure that it is in good condition and of correct size a number of ballast cleaning machines are used. Two different models are illustrated here, both built by Plasser & Theurer. Both types are only suitable for cleaning straight track sections. The machine, above, is of type RM62, and operates with a match wagon, supporting the jib. The machine can deposit spent ballast either in a wagon on an adjacent line or on to the railway bank. The unit weighs 70 tons and is powered by a 235kW engine. The lower illustration is an RM74 machine, which has a self-contained jib (far end) and usually used to deposit ballast into wagons. This type weigh 64 tons in operational condition and powered by a 261kW engine. In 1984 20 RM62 and 22 RM74 machines were in service. *Colin J. Marsden/John Tuffs*

Above and below: One of the most common sights of track machines on the system is the tamper. The two illustrations here are both of the Plasser & Theurer 07-16 Unomatic type. These machines can level, line and tamp, with just one passing on a section of line. They weigh 42 tons and are powered by a 165kW engine. The upper view shows the lining and levelling end whilst the lower plate shows the tamping end.. *Both: Colin J. Marsden*

Above: Two types of Plasser & Theurer levelling, lining and tamping machines are in service — the type 16 and the 32. Both are of similar design, the only difference being the number of sleepers tamped in a single movement. In 1984 there are 80 type 16, and 31 type 32 machines in service. No DX73261, a type 16, is illustrated. *Colin J. Marsden*

Below: This view shows one of the slightly older 07-16 universal machines introduced prior to the types previously illustrated, but the function of the machines are the same. When these units operate on BR main lines a BR driver is provided to act as a conductor to the machine operator, who is responsible for the actual operation, the driver being present to inform the operator of the route and signalling applicable. The photograph shows one of the ER allocated machines at Rotherham during 1983. *Colin J. Marsden*

Above and below: Five type RPM100 machines are in traffic on BR allocated to the LMR, ER and WR and built by Plasser & Theurer. This type is a rail profiling machine with a swarf collector at the rear. The bulk of the body resembles the type 16 or 32 but technically these machines are far apart. The upper view shows the front end of the unit, whilst the lower plate illustrates the swarf collector wagon. These vehicles are numbered DX70101, 79102-5; the swarf collector vehicles carry the same number but with an 'A' suffix. *Both: Colin J. Marsden*

Above and below: Fifteen Plassermatic 07-275 machines are in operation, allocated to all regions. These are a universal levelling, lining and tamping machine for both straight track sections and point/crossing work. Each machine is allocated a material wagon which is semi-permanently coupled to the main machine. From a quick glance these machines again resemble the types 16 and 32, but can be recognised by a low level control cabin positioned between the wheel sets. The upper illustration shows the main machine while the lower shows the material wagon. *Both: Colin J. Marsden*

Above: For use on Scottish, Western and Midland Regions Plasser & Theurer produced three vehicles of the OBW-10 type, titled GP-TRAMM (General Purpose-Track Relaying And Maintenance Machine). These are two part sets, each with a cab and driving position, with the main power equipment housed in the 'A' vehicle which has a larger cab to accommodate additional staff. An Atlas grab is also fitted and the number range of this fleet is 68201-3. *Plasser & Theurer*

Below: Apart from Plasser & Theurer, who are by far the largest suppliers of track maintenance vehicles to BR, some Matisa units are employed, mainly on the ER. This large twin cabbed universal lining, levelling and tamping machine is numbered DB966095 (73501) and was photographed at Knottingley. Matisa built machines are usually larger than Plasser & Theurer units of corresponding type.
Colin J. Marsden

Above: Although Plasser & Theurer are an Austrian company the majority of their machines used in this country are constructed at West Ealing, where a sizeable factory and administration complex exist in the triangle between the WR main line and the Greenford loop lines. Apart from the construction of new vehicles and machines regular maintenance of existing equipment is also undertaken and stock can usually be observed from passing trains. This 1983 view of the main erecting shop shows one of the Atlas grab vehicles under construction. *Colin J. Marsden*

Below: Most BR regions operate at least one Matisa/Neptune track recording trolley. These small four-wheeled vehicles painted in yellow/black livery operate throughout the system producing readouts of track alignment and deficiencies. As well as producing graphs, the trolleys spray coloured dye on defective track enabling the local gangers to establish areas that need attention. These trolleys are numbered in the DX500xx range. No DX50013 is illustrated at Eastleigh in 1981. *Colin J. Marsden*

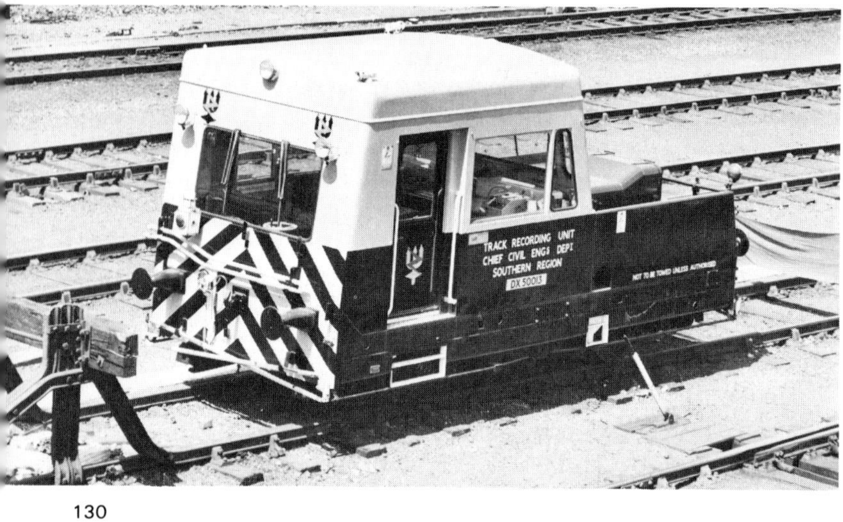

Above and below: Another interesting train to operate on the BR system from time to time is the Speno rail grinding unit, owned by Speno International of Geneva. The five- or seven-coach rail grinding and profiling train operates with a Swiss crew and a British conductor over most main lines on an annual basis. The unit is a self-contained home for its crew for several weeks, with some vehicles designed for staff accommodation with others housing generators. Most vehicles have underslung grinding equipment but not all are used at the same time. The upper plate shows a five car formation looking from the trailer driving end, while the lower illustration is a close up of the locomotive end. The set operates with the No RR555 on each vehicle. Livery is yellow with black underframes.
Both: Tom Noble

Internal User Vehicles

Apart from operating departmental vehicles each region has a fleet of *Internal User* vehicles. These are usually former revenue earning, or departmental vehicles, now retricted from operation on main lines but can be used at a fixed location ie: large marshalling yards, or depots. 'Internal User' vehicles are numbered in a separate range, always starting with 'O' and usually prefixed by the letters IU. The second digit indicates the operating region — 2 LMR, 4 ER, 6, 7 WR, 8 SR, 9 ScR.

Above: Allocated to Waterloo for shunting Waterloo & City stock are two internal user match wagons, these are converts from *Lowfits* and painted in green livery with white lettering. Numbering is IU083405 and 083406, both are vacuum fitted and can usually be found at Waterloo or in the sidings at the Waterloo end of the City line. *Colin J. Marsden*

Below: A coupling problem became apparent at a number of depots and works after *Tightlock* fitted EMU stock entered service, of how to shunt vehicles when they could not be moved under their own power. BREL Horwich, which until late 1983 was responsible for Class 313/507 repairs, overcame this problem by converting a former short wheelbase freight vehicle underframe to a match wagon, with Tightlock/buck-eye couplings at one end and conventional drawgear at the other, no continuous brake being fitted. *Colin J. Marsden*

Above: This ex-revenue earning *Plate* wagon, not fitted with continuous brake, is operating as Internal User vehicle No 041495 in the Sunderland area, and labelled 'Not to leave South Dock'. It was photographed in 1982 at Sunderland depot and the livery was black with white lettering.
Colin J. Marsden

Below: This Internal User vehicle allocated to Colchester is for the transportation of waste oil, and is a steel tank mounted on a former van underframe. It is understood that this vehicle, in common with at least three others, was converted for use with the electrification schemes on the ER(GE) sections some years ago for water transport, and was retained after this and used as an oil carrier for the CM&EE at Colchester depot. *Michael J. Collins*

Left, top to bottom:
This former 12ton *Pipe* wagon, still showing three letter TOPS code ZDV is now in internal use at Guildford where it operates for the power supply section. In 1983 the wagon was used to transfer cable drums from one part of the yard to another. Livery was olive green with white lettering. No external alterations have been made to this vehicle since its revenue earning days.
Colin J. Marsden

Another vehicle now operated by the CM&EE power supply section at Guildford is this former 12ton box van with hinged doors, numbered 083368, used to carry equipment from the Guildford power house to the adjacent yards. The livery is olive green with white lettering, but when photographed in 1982 the wagon was in a delapidated state.
Colin J. Marsden

For internal use at East Wimbledon EMU servicing depot there are two internal user box vans, which were formerly in departmental traffic, however as the wagons were seldom if ever used on the main lines, they were transferred to the Internal User fleet during 1982. Again the livery is olive green with white lettering, this vehicle is numbered 083400.
Colin J. Marsden

For testing diesel locomotive traction equipment at BREL Doncaster this former LMS type van with large sliding doors has been converted into a resistor vehicle and identified as *ZR Resistor No 1*. The wagon now contains some large resistor grids and has roof ventilation doors, accessible by a side ladder. Owing to the additional handrails on the roof this vehicle is not permitted outside the confines of the works, and great care has to be taken even for internal movements as a number of structures could easily be fouled.
Colin J. Marsden

Top right: This former LMS designed CCT-M is now in internal user stock at BREL Doncaster works as a general materials store for the dmu shop, during 1983 it could be found parked adjacent to this shop. This type of vehicle has both side and end doors and were especially suitable for the transportation of motor vehicles. It is now numbered 041345.
Colin J. Marsden

Centre right: This type of vehicle, the BR design CCT, dating from the mid 1950s is now appearing under the internal user flag as much of their designed work has now gone. It is still painted in BR blue livery and now numbered 041644 and used at Whitemoor Yard, March, as a general stores vehicle. *Michael J. Collins*

Below right: There are very few internal user brake vans in service but this former BR standard 20.5 ton vehicle, painted in bauxite, is now in ER internal user service at Diss and labelled 'For Internal Use at Diss'. It is used in association with grain, coal and fertiliser sidings near the station.
Michael J. Collins

Miscellaneous Vehicles and Trains

Above: **PPV.** The Chipmans weed control trains are usually thought of as departmental stock, however these are in fact classified under TOPS as privately owned freight vehicles. All units have been extensively rebuilt to incorporate spray and pumping equipment. Some vehicles are fitted with observation ends, making it possible for them to lead in a train formation, however this is very unusual. The Chipmans weed control train painted in distinctive apple green livery is normally kept at Horsham.
Colin J. Marsden

Left: No, this is not the stock used in the filming of *The Great St Trinians Train Robbery*, but staff transport Wickham type trolleys used on rural lines for staff travelling to engineering sites. These three vehicles photographed at Ferryhill, between Darlington and Durham during 1982 are normally painted yellow with removable canvas sides.
Colin J. Marsden

Below left: This mysterious track borne vehicle is at the Derby Research Centre in the track laboratory, where a complete section of ballasted track exists for testing purposes. The purpose built vehicle is operated over the track under controlled conditions to evaluate various equipment and track wear. The test vehicle has yellow/black warning ends but is not permitted out of the test centre.
British Rail

Right: Equipment carried on some trains operated by the Railway Technical Centre, Derby, is highly complex, sophisticated and in some cases classified. This view inside test car *Mercury* No RDB975280 shows a technician checking performance of some test equipment whilst the vehicle is in motion. *British Rail*

Below: Often Railway Technical Centre trains form a formidable sight around the system, with various weird and wonderful vehicles formed into one train. This formation of Lab No 3 RDB975002, Test car 'Hastings' No RDB975386, POP car RDB975634, APT test car *Pilot*, POP car RDB975635 and LAB No 23 No RDSB975547 is seen in sidings south of Crewe. *John Tuffs*

Above: Prior to the prototype Gas-Turbine APT being constructed two vehicles called 'POP' cars were constructed at Derby for suspension and tilt evaluation. They were originally built as skeleton vehicles but in recent years have had a somewhat conventional bodywork applied. With an APT style articulated bogie nearest the camera supporting one end of POP car No RDB975634, and one end of test car *Pilot*, a tilt/suspension special passes Kidsgrove headed by Class 25 No 25.306 in 1982. *T. R. Moors*

Below: In conjunction with APT design and development work a number of test formations have worked on the WCML route usually over the steeply graded section north of Preston. Two Class 254 power cars provide driving facilities for one such test on 3 August 1982 with a train formed of Mk III test car No 10, an APT match vehicle, an APT power car, an APT match vehicle, two spare IC125 kitchen cars and a TGS. The train was photographed passing Greenholme at the foot of Shap incline. *John Fozard*

Above: Occasionally rare and interesting locomotives visit the Railway Technical Centre, Derby, for tests. During October 1981 SR Class 33 No 33.012 was loaned to the RTC for a test programme and is seen passing Stenson Junction with three research and two Mk I passenger vehicles. Note the front wiring on the locomotive, this would be carried along the side of the locomotive and into one of the test vehicles, enabling recordings of the locomotive's performance to be taken. *John Tuffs*

Below: The Midland main line between Derby and Cricklewood is often host to testing specials, with at least one train per week. This March 1980 illustration shows Class 25 No 25.323 passing Finedon, near Wellingborough, with a south bound test train formed of Test Service car No 7, a Derby laboratory vehicle, and a privately owned tank wagon. This test was made to establish the riding qualities of the PO wagon, information being monitored and logged in the adjoining test car. *Ian Hodson*

Above: The most common departmental special to be seen is the CCE ballast train. This view shows a typical CCE train formed of two *Grampus* type vehicles loaded with sleepers, the brake van is a traffic vehicle classified CAO. The train is headed by a Class 31 and observed at West Ealing.
Colin J. Marsden

Below: This departmental engineers train also headed by a Class 31 is seen departing from Healey Mills yard and formed of a mixture of vehicles all of which form part of a track relaying train. Behind the locomotive is a staff mess van, adjoining this is a tool van followed by a ballast cleaning machine, jib runner, brake van, five rail flats, a *Grampus* wagon and a brake van. *Colin J. Marsden*

Above: When some departmental vehicles are formed into train sets their appearance is rather unusual. This 'train' is the WR tunnel inspection set. Formed behind the locomotive is a mess van with flat roof section, tunnel inspection vehicle — formerly an open service wagon which now has a steel framed platform inside — and a brake van on the rear. *John Tuffs*

Below: The movement of permanent way ballast to engineering sites or from the quarries is usually carried out in 'block' train formation, normally with the same type of vehicle. Here a Class 47 passes Sutton Bonnington hauling a lengthy train of 19 *Seacow* wagons to the stone terminal near Loughborough. After loading the train would be taken to a CCE siding ready to go to an engineering site. *Colin J. Marsden*

Above: This engineers train is formed of a mixture of short wheelbase open wagons. The first vehicle behind the locomotive is a wooden sided open service wagon, followed by a more modern steel bodied wagon of the same type, after this comes a *Grampus*, a Shock-high, folowed by a number of *Grampus*. Class 31 No 31.239 passes Selby on 6 July 1983. *Colin J. Marsden*

Below: Departmental specials need to be arranged when Waterloo & City EMU stock is required for movement to a depot for servicing, as usually the stock is not permitted to operate under its own power on surface lines. This view shows Class 73, No 73.003 hauling a departmental brake van of the SR bogie type and a match wagon either end of a Waterloo & City motor car. *Colin J. Marsden*

Above: Some of the Civil Engineering department crane stock is now getting rather aged, and this seems to especially apply to the WR. This plate shows a Class 47 passing West Ealing hauling a small tool van, crane jib runner and a steam crane, all of GWR origin, with a BR standard 20.5 ton brake van on the rear. *Colin J. Marsden*

Below: Another CCE operated engineers train hauled by SR Class 73 No 73.113 passes St Denys, near Southampton, formed of a well flat for transporting mechanical plant to CCE sites, a former *Railfreight* SPA now under the control of the CCE as a ZZA, followed by three YBA *Sturgeon* rail flats. *Colin J. Marsden*

Above: Approaching Woking from the Guildford line, Class 33 No 33.045 hauls nine YG type *Seacow* hoppers towards the yard to await return to one of the southwest quarries. Often track ballast is moved around the system to pre-assembly depots on a regular time-tabled basis, only operating as 'special services' between pre-assembly depots and engineering sites. *Colin J. Marsden*

Below: The appearance of one of the railway's heavy lifting breakdown cranes is a somewhat daunting sight as usually there has been a rail mishap. Class 33 No 33.012 passes Surbiton hauling the Wimbledon East 75 ton Cowans Sheldon crane, a well wagon carrying a spare bogie, two staff and tool vehicles, a brake van, a weltrol, another tool van, a further weltrol and a brake van. This lengthy train arrived at Surbiton following a serious derailment the previous day when a number of bogies on an EMU formation had been destroyed. *Colin J. Marsden*